THE SCIENTIFIC JOURNAL
Editorial policies and practices

THE SCIENTIFIC JOURNAL

Editorial policies and practices

Guidelines for editors, reviewers, and authors

Lois DeBakey, Ph. D.

In collaboration with

Paul F. Cranefield, M. D., Ph. D.
Ayodhya P. Gupta, Ph. D.
Franz J. Ingelfinger, M. D.
Robert J. Levine, M. D.
Robert H. Moser, M. D.
J. Roger Porter, Ph. D.
F. Peter Woodford, Ph. D.

The C. V. Mosby Company

St. Louis   1976

Copyright © 1976 by The C. V. Mosby Company

All rights reserved. No part of this book may be reproduced
in any manner without written permission of the publisher.

Printed in the United States of America

Distributed in Great Britain by Henry Kimpton, London

Library of Congress Cataloging in Publication Data

DeBakey, Lois.
    The scientific journal: Editorial policies and
practices.

    Includes index.
    1. Technical editing.   2. Journalism, Scientific.
3. Science—Periodicals.   4. Medical writing.   I. Title.
T11.4.D4   070.5′1   76-6046
ISBN 0-8016-1223-3

CB/CB/B   9   8   7   6   5   4   3   2

Participants in discussions

Lois DeBakey, Ph. D., 1968-1975; Chairman, 1971-1975
Professor of Scientific Communication
Baylor College of Medicine, Houston, Texas
Lecturer, Tulane University School of Medicine, New Orleans, Louisiana

F. Peter Woodford, Ph. D., 1968-1973; Chairman, 1968-1971
Executive Editor, *Journal of Lipid Research,* 1963-1969
Managing Editor, *Proceedings of the National Academy of Sciences, USA,*
    1970-1971
Executive Director, Institute for Research into Mental and Multiple Handicap
    London, England

†Ilse Bry, Ph. D., 1968-1972
Editorial Committee, *Mental Health Book Review Index,* 1956-1972
    New York, New York

L. Leon Campbell, Ph. D., 1972-1973
Editor, *Journal of Bacteriology*
University of Delaware, Newark, Delaware

Paul F. Cranefield, M. D., Ph. D., 1968-1975
Editor, *Journal of General Physiology*
The Rockefeller University, New York, New York

George N. Eaves, Ph. D., 1971-1972
President's Biomedical Research Panel
Department of Health, Education, and Welfare, Washington, D. C.

Ayodhya P. Gupta, Ph. D., 1973-1975
Editor-in-Chief, *International Journal of Insect Morphology and Embryology*
Rutgers University, New Brunswick, New Jersey

†Deceased.

Karl F. Heumann, Ph. D., 1968-1973
Executive Editor, *Federation Proceedings*
Federation of American Societies for Experimental Biology
Bethesda, Maryland

Franz J. Ingelfinger, M. D., 1968-1974
Editor, *The New England Journal of Medicine*
Boston, Massachusetts

Robert J. Levine, M. D., 1973-1975
Editor, *Clinical Research*
Yale University School of Medicine, New Haven, Connecticut

Robert H. Moser, M. D., 1974-1975
Editor, *The Journal of the American Medical Association*, 1973-1975
Haiku, Maui, Hawaii

J. Roger Porter, Ph. D., 1973-1975
Editor, *Journal of Bacteriology*, 1951-1962
The University of Iowa College of Medicine, Iowa City, Iowa

William H. Stein, Ph. D., 1968
Formerly Editor, *Journal of Biological Chemistry*
The Rockefeller University, New York, New York

Preface

Although scholarly journals have been in existence since January 5, 1665, when *Le journal des sçavans* first appeared in France, editors have had little guidance in establishing their policies and practices or in resolving difficult problems that arise in the course of their work. Each has had to struggle with these problems individually, sometimes without being fully aware of the advantages and disadvantages of a specific decision. Recognizing the need for editorial guidelines, Dr. F. Peter Woodford invited a group of his colleagues to join him in surveying activities related to the editing of a scholarly scientific journal. The original group, formed in 1968 as the Committee on Editorial Policy of the Council of Biology Editors, consisted of Dr. Woodford as Chairman, the late Dr. Ilse Bry, Dr. Paul F. Cranefield, Dr. Lois DeBakey, Dr. Karl F. Heumann, Dr. Franz J. Ingelfinger, and Dr. William H. Stein. When Dr. Woodford moved to England in 1971, I succeeded him as Chairman. Later Dr. Ayodhya P. Gupta, Dr. Robert J. Levine, Dr. Robert H. Moser, and Dr. J. Roger Porter joined the group, which continued its activities, ultimately, on an independent basis.

The group met several times a year to discuss and debate editorial problems, and had voluminous correspondence between meetings. It was not, however, the charge, or original intent, of the group to publish a book; that decision came later in response to an expressed need. When I became Chairman, I wrote a large number of journal editors to solicit their views about additional topics they would like to have us explore. Their responses, along with numerous requests I received from throughout this country and abroad, convinced me that we should make the results of our deliberations available in the form of a book. I made the proposal to those actively participating in the discussions at the time, and they concurred. We subsequently re-examined statements we had previously prepared, and recast, or completely rewrote, them for incorporation with the material expressly prepared for this book.

Although most of the collaborators edit biological or medical journals, many of the policies and practices discussed apply to scholarly journals of any kind. Because we are aware that there is no "right" policy on most editorial matters, we have tried not to prescribe rules, but have, instead, explored various facets of the problems that confront the editor in his daily work. For convenience, we have divided the book into two general sections: editorial policies, which usually require major decisions; and editorial practices, which involve minor decisions, often about format or mechanical style. In some instances, we have made recommendations, whereas in others, when recommendations were considered beyond our combined experience and knowledge or when issues are in a state of extreme flux, we have presented the results of our deliberations in the form of discussions. In any group of independent thinkers it is sometimes difficult to arrive at a consensus; when full agreement could not be reached, we have abided by majority opinion.

Having probed the issues seriously for several years, we are acutely aware of the difficulties that guidelines inevitably raise. We have found it necessary, for example, in the course of our discussions and deliberations, to revise various chapters several times as the current status of, or attitude toward, those issues has changed. We therefore claim no permanence for our current views and no pretense at completeness. We hope, however, that the present treatment will be of some benefit to other editors and will stimulate further exploration of the subject. With this purpose in mind, we have chosen to publish our essays in their present form rather than continue to postpone their publication pending further analysis and refinement.

That certain statements seem self-evident is unavoidable if a book of this kind is to be of practical value to both novice and experienced editors and reviewers. Some of the information may be of interest to authors and readers of journal articles, both of whom may wish to know what happens to manuscripts after they are submitted for publication and how decisions are made regarding acceptance or rejection.

We are indebted, first, to Dr. Woodford, who conceived the idea of examining the editorial policies of professional scientific journals and guided us through some confused and even stormy early meetings when we were defining our purpose. It saddens us that Dr. Bry, whose penetrating comments helped keep us alert to opposing arguments and views, did not live to see the culmination of our deliberations in this book. We are grateful to many editors of scientific journals who responded to requests for information about their editorial policies and problems and to others who attended our meetings as guests. Their contributions have been useful in broadening our view of the issues. Appreciation is also due Dr. L. Leon Campbell and Dr. George N. Eaves, who attended some of our meetings and participated usefully in the discussions. I owe personal thanks to each of my collaborators, who took time from other pressing duties to participate in this endeavor; a more stimulating, dedicated, and concerned group would be hard to find. Working with them was not only rewarding, but enjoyable. Special gratitude is extended to Selma DeBakey for her patient, critical reading of the successive drafts of this manuscript throughout its preparation.

Lois DeBakey

Contents

ix

APPENDICES

THE SCIENTIFIC JOURNAL

Editorial policies and practices

CHAPTER 1

The purpose of scientific journals

In much of the discussion in this book, we take for granted certain attitudes toward science and scientific publication. We assume that the purpose of a scientific journal is to facilitate communication among scientists and that the goal of scientists is the discovery of scientific knowledge and the verification of such discovery. We thus assume that the purpose of a primary research journal is to publish results of scientific investigation that have already been proved to be valid and of enough importance and interest to warrant the expense of publication. This leads to the simple rule that a journal should publish what is *new, true,* and *important.* But how does one decide if something is new, true, or important? And what conflicts arise in the attempt to apply these criteria?

The conflict between discovery and verification, that is, the conflict between *new* and *true,* is by far the most difficult to resolve. The scientist who believes that he has made an important new discovery or has an important new insight is anxious to share his discovery or insight quickly with the world, so that he may receive credit for his originality. He may, therefore, be impatient when reviewers or editors are not convinced that his new knowledge is true, that is, that he has verified his discovery. It is easy to assert that the editors and reviewers of the journal have the right and responsibility to demand the highest possible standard of verification as a condition for publishing an article. But one cannot say whether any statement is true; he can only say that he is or is not convinced by the arguments offered in its favor. In a certain sense, therefore, a balance is always struck: if a theory or discovery seems reasonably well proved, if it is new and seems likely to be important or provocative and suggestive, the balance will tip in favor of publication.

Can any harm result from holding to an absolute standard of truth? Yes, considerable harm can result. If an idea that is new, important, provocative and suggestive, and *true* is denied publication on the grounds that it is not well enough established, or proved, or verified, the result may be a serious loss to the progress of scientific knowledge. On the other hand, the fact that a would-be author is convinced of the originality and validity of what he wishes to publish is not enough to satisfy most editors. No editor would knowingly publish assertions that are false; no editor would willingly decline to publish something new and important solely because it had not yet been fully proved. But is the line easy to draw?

Are there articles that are true, but neither *new* nor *important,* that should be published? Perhaps not, yet some articles containing information that is neither wholly new nor very important may well be published because they are so completely true. A comprehensive survey of previous publications on the subject, coupled with a careful reinvestigation of the problem, followed by a set of factual statements so thoughtfully considered and documented that they seem to settle a question once and for all, may well be worth publishing even if the conclusion is much the same as was always suspected and even if the problem thus settled is not of prime importance. It is, after all, rare in science that anything is finally settled.

What about material that is true and important, but *not* wholly *new?* Although few editors would care to publish the results of research that only duplicate results obtained years before, it is not uncommon for them to publish confirmatory data obtained with new or more precise methods. In addition, facts previously established but almost forgotten may be republished because they have suddenly become much more important. The novelty of the findings is thus not absolute, but is judged in relation to a newly perceived importance. Finally, an article may appear that contains observations previously reported, but forgotten by all concerned—author, reviewer, and editor—so that facts that are not at all new may inadvertently be published. Such unintentional "rediscovery" may, however, be very important and open new avenues of highly productive research, while the previous finding may continue to lie unnoticed until someone familiar with the new work happens to come upon the old. The demand that material in an article be new is not, therefore, one that can be enforced in a rigid and automatic way.

What about the article containing data that are clearly new and clearly valid but seemingly *not* very *important?* Again, questions of judgment arise. The full importance of a given fact or theory is not always evident at the time of the discovery or of the advancement of the theory, and the mere novelty and solidity of an assertion may therefore lead to publication even though the importance is not particularly obvious.

So although the purpose of a primary research journal may well be the publishing of scientific information that is new, true, and important, it is not easy to establish these criteria. Few published articles, in fact, meet absolute standards of novelty, proof, and significance. Furthermore, harm may be done both to the investigator and to the progress of science by the failure to publish an article merely because it falls short when judged rigidly by those three criteria. A more reasonable decision will rest on a balance among the three criteria.

If a journal has as its purpose the publication of what is new, true, and important, the motive for such publication must be the availability of the material to potential readers. Vexatious problems thus arise about organization and style, about jargon and abbreviations, about citations and documentation, about the quality of tables and illustrations, about statistics, and about reproducibility. All these matters have some bearing on the presentation of the evidence, which enables the editors and reviewers to decide that the material is new, true, and important, but once that decision has been reached, the question arises of whether the material has been presented in a way to reach a wider audience than the dozen or so people currently working on a closely related problem. Because editors hope that the articles published in their journals will be read in many countries and that at least some of the articles will be read in later years, they discourage local idiom and laboratory jargon. Editors also hope that at least some of the articles in their journals will be comprehensible to persons in related fields, and so they try to persuade authors to place the work in some sort of context by means of an introduction and discussion.

Some authors contend that if their work is new, true, and important, it is hardly fair to ask for *comprehensibility* as well, maintaining that questions of organization and style are often subjective. Nonetheless, most editors will hope that articles in their journals have all four qualities: new, true, important, and comprehensible. Since articles often fail to meet all these criteria, the editor may wish to provide mechanisms by which articles can be challenged; by which claims to novelty or to truth can be attacked; by which errors can be corrected in later articles; by which obscurely presented concepts can be made clear. Not all these functions necessarily belong to a primary research journal; some may belong, for example, to the review journal.

If we grant, then, that the purpose of science is discovery coupled with verification, that is, discovery critically examined to ensure that what appears to be discovery really is discovery, then the purpose of a primary research journal is to communicate such verified discovery, and its criteria will probably be that the contents of an article shall be *new, true, important, and comprehensible*. And almost everything that is discussed in this book deals in some way with those criteria and their application.

CHAPTER 2

The editor: role, functions, responsibilities, and administrative arrangements

The administrative responsibility for a journal is generally divided among three persons—the *owner*, the *publisher*, and the *editor*. The same person may occupy more than one of these positions. The owner defines the authority of the editor as well as of others with whom the editor may be expected to consult. Since readers usually hold the editor accountable for all aspects of a journal and since every part of a journal reflects on the reputation of the editor, he will find it advisable to participate as actively as possible in policy decisions regarding not only the intellectual content of the journal, but its production as well, including such matters as advertisements, make-up, sequence, typeface, and cover. Even though the editor may delegate most of the responsibility for the mechanics of production to others, unless he monitors these activities, his editorial reputation may be placed at risk by decisions over which he has no jurisdiction or control.

OWNERSHIP OF THE JOURNAL

The owner of the journal ordinarily governs the general policies of the journal in such matters as content, format, and finances. Some journals are owned by professional societies, and others are owned by independent commercial enterprises. A common variant is the *commercially-owned*, but *society-edited* journal. When a professional society owns the journal, it may either contract with a publisher, or, in the case of some large organizations, it may publish the journal itself.

Long-range policy decisions for society-owned journals are ordinarily made by a committee appointed by the society, commonly called the publications committee. Commercially-owned journals have their policies established through any of a large variety of administrative arrangements. Commercially-owned, but society-edited journals may also have policy decisions made by publications committees or committees on editorial policy.

The editor should have a clear understanding with the owners about the type of material that may be included in the journal without the editor's review or approval. In the case of commercially-owned journals, for example, the editor may not have the prerogative of reviewing advertisements or statements written by the publisher. In the case of society-owned journals, the editor may be obliged to publish proceedings of scientific and business meetings, abstracts, committee reports, and the like. Since the editor will be held accountable by readers for all contents of the journal, he should consider, in advance, the potential for his being embarrassed by the publication of material that he has not had an opportunity to review.

EDITORIAL BOARD

The editorial board is a panel of advisers whom the editor may consult for various purposes. In some instances, the editorial board may evaluate manuscripts submitted for publication. In other cases, the editorial board may simply suggest the names of competent reviewers for specific manuscripts, without themselves acting as reviewers. In still other cases, the members of the editorial board function as *section editors,* with responsibility for journal coverage of a particular specialty, such as cardiology, geriatrics, immunology, microbiology. They may solicit material for the journal, write editorials themselves, or otherwise see that the journal properly reflects activities, and the state of knowledge, in their particular disciplines. Some editorial boards also serve the function previously specified for the publications committee, that is, the establishment of policy. It may be preferable to keep these functions separate.

The editor should have the authority to initiate nominations for membership on the editorial board, which can then be approved or disapproved by the policy-making group. Selection of the editorial board requires consideration of several factors. The temptation is strong to recruit distinguished "names" in the field to enhance the prestige of the journal, but such distinguished "names" are not always competent reviewers and are not necessarily cooperative. The editor would do better to select those whom he knows to be willing and able to provide sound advice and constructive criticism. The editorial board should contain sufficient breadth of expertise to accommodate the various subspecialties within the general field covered by the journal.

The persons named to the editorial board should be those whom the editor consults most frequently for review of submitted manuscripts. For one thing, when an editor asks for someone to review a manuscript, he is, in effect, asking for several hours of that person's time. Although some journals provide token reimbursement, virtually no journal has the budget to compensate the reviewer fully. The service rendered to the journal and to the discipline by a frequently consulted reviewer should be recognized by the inclusion of his name in the membership of the editorial board. Moreover, authors who are considering submitting manuscripts to a journal like to know what types of reviewers are likely to be judging and criticizing their papers. The published list of names on the editorial board also signals to prospective authors the range of interests covered by the journal.

Some mechanism should be established to add new members to, or remove old members from, the editorial board. Requirements for new members will usually be determined by needs to accommodate advancing scientific knowledge and technology or to replace consultants who, for one reason or another, have become inactive. Ordinarily, rotation is accomplished by appointment of members to the editorial board for a specified term. The editor should have the prerogative of terminating any appointment to the editorial board of a member who proves ineffective or uncooperative. Since, however, terminating an appointment may be difficult, or at least awkward, it is desirable to avoid making the initial appointment protracted; an initial appointment of one or two years will allow the editor to make a fair judgment of the member's performance. The services of a particularly valuable consultant can usually be retained by renewal of his appointment.

DELEGATION OF RESPONSIBILITY

Most editors will find it necessary to delegate responsibility for some decision-making. Since the editor remains responsible for decisions made by his delegates, he should choose them carefully. Delegation of responsibility can be divided into two major categories: (1) editorial functions and (2) administrative functions.

(1) Editorial functions

The persons to whom editorial responsibilities are delegated are usually called *associate editors*. Associate editors may be assigned responsibility on the basis of their subspecialty interests, in which case they may be called *field editors*. A journal may, for example, have a field editor for internal medicine, pediatrics, general surgery, genetics, or plant pathology. Each field editor will have a set of subspecialty advisers with whom he consults. Large journals may have associate editors responsible for separate departments, such as original articles, correspondence, abstracts, book reviews, and news. International journals may have *regional editors,* each of whom is assigned responsibility for handling manuscripts from the hemisphere or continent in which he resides. Regional editors may, in turn, have associate editors or field editors to assist them.

(2) Administrative functions

Administrative functions may be similarly delegated. Depending upon the needs and budget of the journal, the editorial staff may include a managing editor, administrative assistant to the editor, or secretary, separately or in combination. To these assistants, according to their capabilities, may be assigned responsibilities for handling editorial correspondence, keeping records of the status and location of manuscripts, reminding reviewers when they are late in returning manuscripts, supervising redactors, indexing, managing the budget of the editorial office, and the like.

The terminology used to identify various positions on the editorial staff varies widely. The editor should be cautioned, in considering recruitment of new staff members, that the previous title held by a prospective employee may provide little indication of his capabilities or experience. Persons who have precisely the same responsibilities may be variously designated: assistant editor, assistant to the editor, editorial assistant, managing editor, executive editor, administrative assistant, secretary. The editor is advised to obtain as precise as possible a description of the prospective recruit's previous assignments and capabilities. One particularly ambiguous designation is that of managing editor: in some cases persons with this title serve as secretaries, whereas in others they function in responsible administrative roles, handling most aspects of the transactions between the editorial office and the journal's owner, publications committee, authors, readers, indexers, advertisers, and others. In still other cases, managing or executive editors may be assigned decision-making responsibility regarding acceptance of various types of submitted material, including, at times, original articles.

For a sample organizational chart of a journal editorial office, see page 111.

EDITORIAL POLICIES

Chapters 3-17

Reviewing

Chapters 3-7

Special types of manuscripts

Chapters 12-17

Reviewing

CHAPTER 3

The manuscript reviewing system

The system described here is designed for those editors who intend to use reviewers to assist them in making editorial decisions. We shall not discuss the advantages and disadvantages of the reviewing process as such. Nor shall we discuss every possible aspect of editorial policy related to reviewing, although we hope to have made a useful start on some of the leading problems.

In this section, each question that the newly-appointed editor will need to ask himself is followed by a suggested answer, which is then discussed, with pros and cons where appropriate, and with consideration of special cases. The discussions are intended to be thought-provoking, and not in any way to abrogate the right of every editor to make up his own mind on each question presented.

The term "reviewer" is used throughout to denote any expert adviser to the editor; no distinction is drawn between "outside" reviewers and members of the editorial board or others named on the masthead of the journal.

(1) *What is the optimum number of reviewers for each manuscript?*
Most editors select two reviewers per manuscript. A third reviewer, or even more reviewers, may be necessary in special cases and in instances in which the first two reviewers disagree.

DISCUSSION

For a journal (or section of a journal) of narrow scope, it can plausibly be argued that a single editor is capable of making acceptance-rejection decisions without asking for advice from outside reviewers. Most editors believe, however, that it is wise to have at least one other opinion as a check on their own judgment. (A few journals operate on the principle that the editor alone may accept a manuscript, but that two concurring opinions are needed for rejection.) When the editor is well-informed on the subject of the article, one other reviewer may well be sufficient. When he is not,

11

two reviewers seem desirable, so that they may act as a check on each other. When the paper in question has several aspects demanding different kinds of expertise, additional reviewers may be needed. It is not wise to increase the number of reviewers indefinitely, because this inevitably lengthens the reviewing process (which seems unfair to the author, however weighty the final accumulation of comment may be) and increases the load on the limited number of competent reviewers available.

In general, then, the most practical number of reviewers for each manuscript is probably two, with the addition of others in special cases.

When the two reviewers give contradictory advice, the editor has the following options. He can (a) himself arbitrate between the reviewers (preferably after discussion by telephone), (b) ask a third reviewer to arbitrate, or (c) ask for one or more fresh reviews without reference to the first two. The third possibility is often supposed to be fairer to the author, but often lengthens the reviewing process without materially facilitating a decision.

(2) *Should reviewers be identified to the author?*

On balance, the advantages of preserving a reviewer's anonymity, and thus protecting him from pressures that might lead him to modify or distort his real opinion of a manuscript, outweigh the main disadvantage, which is that the reviewer shielded by anonymity may indulge in irresponsible criticism.

DISCUSSION

The leading argument in favor of anonymity is that it permits a reviewer to be frank without becoming embroiled in personal animosities. In general, anonymous reviewing is freer, more candid, and therefore more useful both to the journal and to the author.

The chief risk of anonymity is that it may be used as a shield for prejudice or unfair comment, but an editor can guard against this in several ways. By securing at least two reviews of any article, he should ensure that every review is to some extent checked by another. When two reviews are widely discrepant, and especially when the editor suspects prejudice from the tone of one of them, he can send the reviews and the article to a carefully selected third reviewer.

One undesirable consequence of signed reviews is that the author may attempt to resolve any questions by communicating directly with the reviewers instead of with the editor. Since the editor must make the final decision, it can be embarrassing for him to be presented with a revised manuscript accompanied by a statement that all the problems have been ironed out between the author and the reviewers, and that there is presumably now no impediment to publication.

Most reviewers prefer to remain anonymous, and there is no valid reason not to honor that preference. On the other hand, some reviewers prefer to identify themselves. If a reviewer so chooses, he may sign his name to the copy of his review that is to be sent to the author; the editor then has little choice but to permit him to do so, or lose his services. It seems reasonable for an editor who wishes to rely for the most part upon anonymous reviewing to permit reviewers to identify themselves, without encouraging them to do so. The editor who receives a signed review may wish to ask the reviewer if the signing was deliberate before he passes the signed review on to the author.

(3) *Should the author's identity be concealed from the reviewer?*
Although, theoretically, a review might be more impartial if the reviewers are unaware of the author's identity, removal of the author's name from the title page of a manuscript sent out for review seems futile.

DISCUSSION
Authors complain that a reviewer is subject to two kinds of bias when he sees the author's name and affiliation on a manuscript: (a) he may be unduly influenced by the reputation of the author or his institution, or (b) he may be in competition with the author, and may try to delay publication of the paper by criticizing it unnecessarily harshly.

Removal of the author's name from the title page will not circumvent these biases. The manuscript itself usually provides sufficient internal evidence to disclose its source, and the reviewer's attention is, if anything, directed more to the authorship than it would otherwise be. Part of an editor's responsibility is to be alert to the effects of bias, to distinguish between documented and subjective, unsupported criticism, and to discontinue consulting reviewers who repeatedly evince bias.

(4) *Should a reviewer with known prejudice against an author be asked to review his manuscript?*
In general, if a reviewer is known to be antagonistic to a particular author, it is not advisable to consult him about that author's manuscript.

DISCUSSION
Some people claim that, if the prejudiced reviewer is otherwise the best-qualified expert on the subject of the article, the editor may well invite his comments if he is prepared to go to the trouble of weeding out *ad hominem* criticism in order to obtain specialized insight. The arguments against this approach are: the trouble is often considerable; the editor must risk antagonizing the reviewer if he decides to overrule his advice (of which

there is a higher possibility than with a neutral reviewer) ; and even when the editor has reduced the criticisms to what he thinks are substantive ones, he may be left with a harsher review than other manuscripts received.

(5) *Are reviewers generally paid?*
Most reviewers regard the evaluation of manuscripts as a professional responsibility and an honor, for which they will receive no remuneration.

DISCUSSION
If an editor uses a reviewer particularly frequently, he may decide that the journal should at least defray postal and secretarial expenses, and should perhaps consider making a nominal payment in recognition of the reviewer's services. If the payment were not nominal but truly consonant with the reviewer's qualifications and the amount of time spent, the expense would be so great that the whole economic structure of journal publication would have to be revised.

(6) *How much guidance should the editor give to reviewers?*
Elaborate and detailed instructions may be self-defeating, but reviewers consulted for the first time are grateful for some general guidelines and for convenient forms on which to submit their evaluations and recommendations.

DISCUSSION
Many editors use printed Guidelines for Reviewers, whereas others make a practice of consulting only experienced reviewers, who do not need such guidance. An editor who considers such counseling advisable may want to use the Suggested Guidelines for Reviewers in Chapter 4.

(7) *Should reviewers be asked to grade manuscripts?*
Some form of grading may be helpful to the editor, but if used it should be restricted to a small number of grades; its purpose and the criteria to be applied should be fully explained.

DISCUSSION
An overall "grade" for a manuscript may seem simple-minded, naive, and unscholarly to some reviewers, and for this reason should not be obligatory. On the other hand, it may bring some reviewers' recommendations into focus (the grade is often surprisingly at variance with the tone of the specific comments). The grade should, of course, be given separately from the comments intended for the author. A simple system is the most useful: A (outstanding), potentially A after revision, B (good), potentially B after revision, C (fair), D (substandard).

Alternatively (or additionally), an editor may bring the recommendations into focus by asking the reviewer to comment confidentially on specific aspects of a paper, such as the following (to be appropriately modified for each journal, and to be considered by the reviewer insofar as they apply to the particular paper) :

importance of research question or subject-field studied,
originality of work,
appropriateness of approach or experimental design,
adequacy of experimental techniques,
soundness of conclusions and interpretation,
relevance of discussion,
clarity of writing and soundness of organization of the paper.

Specification of the kinds of comments desired aids not only the inexperienced reviewer but also the experienced reviewer who is competent to make these judgments, but who may overlook them because he has become lost in detail.

(8) *Should reviewers be allowed or encouraged to pass manuscripts to colleagues for review?*
This privilege may be extended to trusted and frequently consulted reviewers, with the proviso that the actual reviewer's name be given to the editor and that the originally chosen reviewer ensure that his colleague understands the confidential nature of the reviewing process.

DISCUSSION
Review by junior members of a department can be excellent: more detailed and less conservative than that of a senior, overcommitted consultant. If the senior man is conscientious in discussing the review with his junior colleague, the colleague receives valuable training. It is, however, unwise to grant permission, indiscriminately, to all reviewers to use subreviewers. Unless such permission is specifically granted, the editor assumes that the originally chosen reviewer reviewed the paper.

(9) *Should reviewers' detailed comments be sent to the author?*
When a reviewer supplies detailed comments on a paper, these should usually be sent to the author together with the editor's decision letter.

DISCUSSION
When the editor requests revision of the manuscript, the reviewers' comments presumably contain at least some of the suggestions on which the revision is to be based and therefore should be sent to the author. The editor may wish to indicate which comments he considers especially important.

When the editor decides to reject a manuscript, he may hesitate to send the comments because (a) they were favorable, and therefore contrary to the decision based on another reviewer's comments or on other considerations, or (b) they may be interpreted as an invitation to resubmit the article with a rebuttal of the comments. With both these considerations in mind, an editor may be tempted to withhold the reviewers' comments when he sends a letter of rejection.

We believe that this attitude is improper. Suitably worded rejection letters leave no doubt of the editor's position and of his reasons for sending the comments. For example, he may reject an article, giving no reason other than that there is no space, the overall rating is not high enough, or the subject of the article is not in scope. He may also enclose the reviewers' comments, some of which are favorable, and append a remark such as "I am enclosing the comments made by the reviewers, since they may be useful to you should you decide to revise the article before sending it to another journal." This statement does not imply that the reasons for rejection are given in the reviewers' comments. Any remark of a reviewer's that specifically recommends acceptance of the manuscript should be deleted, merely to prevent confusion. Alternatively, the editor can summarize his reasons for rejecting the article and allude to the reviewers' comments for details. He can clearly state that he is not inviting the author to resubmit the article after revision in the light of the comments. It is not good policy, however, for an editor to be overly eager to squelch authors' rebuttals, since neither editors nor reviewers are infallible.

The main point is that the editor who refrains from sending the comments, or an edited version of them (see item 10), is depriving the author of valuable criticism, of a kind he may not obtain from his close colleagues. If the article has been rejected, the author need not accept the criticism unless he wants to, but it does not seem fair to reviewer or author to "waste" the comments.

A further consideration is that if the article is submitted to another journal, the same reviewer may be asked to judge it again. He may assume that the author has seen his comments, and when he finds no changes made in response to any of them, he may become antagonistic to the author. If, on the other hand, he divines that the editor did not pass his comments along, he may become annoyed with the editor. In either instance, he may feel that he has wasted his time in making suggestions, which he assumed would have some beneficial effect on a future version of the article, even though it might be destined for another journal.

(10) *If reviewers' comments are sent to the author, should they be edited or kept inviolate?*
Reviewers' comments should be edited on occasion, in order to remove offensive wording but not to suppress criticism.

DISCUSSION

A reviewer may express a well-justified criticism in terms that are too acidulous for an author's digestion, and it is the editor's responsibility to watch for such terms and modify them. It is not always easy to judge what an author will find offensive, and it is, unfortunately, easy to edit out the thrust of a criticism as one softens it. One can sometimes avoid rewording the comments by inserting a placatory sentence in the decision letter to the effect that the comments, though perhaps strongly worded, may nevertheless prove useful. Often, wording which seems offensive was not so intended by the reviewer; a tactful note to the reviewer can improve his later performance in this regard.

Any remark in the reviewer's comments that is contrary to the editor's decision about the acceptability of the manuscript should be deleted, because it leads to confusion and because the decision is that of the editor, not of the reviewer.

(11) *Should reviewers be told whether the manuscript was accepted?*
A reviewer appreciates being informed of the ultimate disposition of a manuscript that he evaluated.

DISCUSSION

Informing the reviewer of the final decision is common courtesy and is easily achieved by sending him a copy of the "decision letter" if time does not allow a personal letter. When the reviewer's advice has been overruled, the editor should add a note of explanation. The trouble is slight; the rewards are great.

(12) *Should a reviewer see the comments of any other reviewer(s)?*
When the editor has accepted, rejected, or requested revision of a manuscript, he should send each reviewer a copy of the comments of the other reviewer(s), together with the decision letter.

DISCUSSION

This procedure, which is neither troublesome nor expensive, offers certain benefits. Not only does the reviewer learn what another reviewer thought of the manuscript, but the editor may learn from reviewer number two that reviewer number one actually misinterpreted some part of the manuscript.

If the editor disagrees with one review or is skeptical of its validity, he can append a note to that effect as a point of information for the reviewer who receives it.

The anonymity of the reviewers is best preserved in this transaction, partly to prevent the development of hostility between reviewers, partly to assure all concerned that anonymity in all transactions, including those with authors, is scrupulously preserved.

Reviewing

CHAPTER 4

Guidelines for reviewers

Whether a journal editor is well advised to have a printed set of Guide-lines for Reviewers is debatable. Many editors attach guidelines to each manuscript to be reviewed. Some reviewers welcome them, especially if they review manuscripts for several journals with different editorial policies. Other reviewers called on frequently by the same journal may feel con-strained by being instructed in the conduct of a scholarly activity for which they have developed their own philosophy and procedures.

For the editor who has decided that the advantages outweigh the dis-advantages, we offer some Suggested Guidelines for Reviewers, with a dis-cussion of each, to assist the editor in formulating his own. Each suggested guideline, which is in *italics,* is directed to the reviewer and phrased in such a way that it can be incorporated without change in a printed set of guidelines. The discussion that follows the suggested guideline is addressed to the editor, not to the reviewer, and is not intended as part of any printed set of guidelines to be sent to reviewers.

SUGGESTED GUIDELINES FOR REVIEWERS

(1) *The unpublished manuscript is a privileged document. Please protect it from any form of exploitation. Reviewers are expected not to cite a manuscript or refer to the work it describes before it has been pub-lished, and to refrain from using the information it contains for the advancement of their own research.*

DISCUSSION

Although the content of this guideline may seem obvious, it is advisable to remind the reviewer that the manuscript is being sent to him solely in order that he may advise the editor as to its acceptability. The last phrase perhaps contains a counsel of perfection, but it is an ideal that must be striven for if the review system is not to fall into disrepute.

(2) *A reviewer should consciously adopt a positive, impartial attitude toward the manuscript under review. Your position should be that of*

18

the author's ally, with the aim of promoting effective and accurate scientific communication.

DISCUSSION

This guideline may counteract any inclination for a reviewer to act like a censor or hostile advocate, with an obligation to find fault with submitted manuscripts.

(3) *If you believe that you cannot judge a given article impartially, please return the manuscript immediately to the editor, with that explanation.*

(4) *Reviews should be completed expeditiously, within* (state here the time you consider reasonable, for example, two weeks). *If you know that you cannot finish the review within the time specified, please telephone the editor (collect) to determine what action should be taken.*

DISCUSSION

Editors may find it helpful to attach a sticker to the envelope containing a manuscript sent for review, which asks the recipient (secretary or receptionist) to telephone the editor if the reviewer is out of town. This is, of course, unnecessary if the editor follows the practice of telephoning the reviewer beforehand to ascertain his availability.

(5) *A reviewer should not discuss a paper with its author.*

DISCUSSION

For inexperienced reviewers, it seems natural and reasonable to discuss points of difficulty or disagreement directly with the author, especially if the reviewer is in favor of publication and does not mind revealing his identity. Specific prohibition of the practice is therefore necessary, because the other reviewer and the editor may have differing opinions, and the author may be led into undue optimism by having "cleared things up" with the reviewer who made contact with him directly.

(6) *Please do not make any specific statement about the acceptability of a paper in your comments for transmission to the author, but advise the editor on this score either in a confidential covering letter with your comments or on the form(s) provided for that purpose.*

DISCUSSION

The responsibility for acceptance or rejection must lie with the editor, who will base the decision not only on advice from reviewers but also on considerations outside the reviewers' purview. Statements concerning acceptability in a reviewer's comments may, therefore, run counter to the editor's decision and may have to be deleted. The guideline is intended to avoid this troublesome necessity. The editor should see to it that application of the guideline does not result in transmission to the author of seemingly curt reviewers' comments, which do not state that the reviewer enjoyed or admired the paper but may present several minor, adverse criticisms. In his letters to authors, the editor should make amends for any lack of graciousness in the reviewers' comments.

(7) *In your review, please consider the following aspects of the manuscript, as far as they are applicable:*
> *importance of the question or subject studied,*
> *originality of the work,*
> *appropriateness of approach or experimental design,*
> *adequacy of experimental techniques,*
> *soundness of conclusions and interpretation,*
> *relevance of discussion,*
> *clarity of writing and soundness of organization of the paper.*

DISCUSSION

Specification of the kinds of comments desired aids the inexperienced reviewer as well as the experienced reviewer who is competent to make these judgments, but who may overlook them because he has become lost in detail. Editors will, of course, select from or modify the above list according to the needs of their journals.

(8) *In comments intended for the author's eyes, criticism should be presented dispassionately, and abrasive remarks avoided.*

(9) *Suggested revisions should be couched as such, and not expressed as conditions of acceptance. In a separate letter to the editor, please distinguish between revisions considered essential and those judged merely desirable.*

DISCUSSION

Although the reviewer may advise the editor that he considers some revisions essential for acceptability, he should not so state to the author because (a) the editor may not agree and (b) the author may mistakenly conclude that adoption of the revisions named will guarantee acceptance of the revised manuscript. If the list of suggested revisions is long, both editor and author will be grateful for a distinction among important, less important, and minor ones.

(10) *Your criticisms, arguments, and suggestions concerning the paper will be most useful to the editor if they are carefully documented.*

DISCUSSION

Reviewers sometimes make dogmatic, dismissive statements, particularly about the novelty of work presented in a manuscript under review, which under strict examination prove to be unjustified. It seems only fair to the author, therefore, to ask reviewers to substantiate their statements. However, some editors may feel obliged not to apply this guideline on the grounds that reviewers are already so overworked that this requirement would make it impossible for many of them to act as reviewers at all. An editor who takes this stand must be especially careful to consider sympathetically an author's rebuttal of reviewers' comments.

(11) *You are not requested to correct deficiencies of style or mistakes in grammar, but any help you can offer to the editor in this regard will be appreciated.*

DISCUSSION

Reviewers like to be told whether this kind of help is expected or desired. Many editors do not desire such help. Those who do might expand the guideline with details of the kind of help desired (such as drawing attention to unclear or ambiguous passages or sentences, suggesting reorganization, or pointing out the need for condensation of particular passages). The editor may also want to specify that if a reviewer wishes to mark the text of the manuscript, he should either use a very soft pencil or else make a photocopy, mark it, and return it together with the original. If the editor decides to reject the paper, he will usually not want to return to the author a manuscript that has been heavily marked.

(12) *A reviewer's recommendations are gratefully received by the editor, but since editorial decisions are usually based on evaluations derived from several sources, a reviewer should not expect the editor to honor his every recommendation.*

DISCUSSION

This guideline is intended to forestall, at least partly, the sense of disillusionment the reviewer feels when the editor does not follow his recommendations to the letter.

CHAPTER 5

Duplicate reviewing

A reviewer who has evaluated a given manuscript for one journal is sometimes asked to review the same manuscript, or a revision thereof, for another journal when the editor of the first journal has rejected it. If the reviewer's first evaluation was favorable, he should still re-examine the manuscript for possible changes and for its suitability for the second journal. If the reviewer's first evaluation was negative, several courses of action are possible:

(1) The reviewer can again review the manuscript. If it has been revised satisfactorily in response to his earlier criticisms, he may simply submit a favorable report. If, however, he again submits a negative evaluation, the author has, in a sense, been placed in double jeopardy; he has, in asking for a new trial, been given the same judge. If, as is often the case, the author and the reviewer disagree on some fundamental issue or aspect of method, the author's opportunities to publish his case are compromised.

If the initial negative evaluation was capricious, prejudiced, or erroneous, one may argue that not only the author but scientific knowledge in general may be harmed by permitting the same reviewer again to influence editorial decision. If, on the other hand, the initial negative evaluation was objective and the author has ignored the sound advice and criticism offered in the prior review, it is disadvantageous to the editor not to have this information.

If the reviewer again reviews the manuscript, it is desirable that he inform the editor that he has seen it before. The editor may, in fact, wish to request that reviewers do so. This will alert the editor to the possibility that the review might be prejudiced and will enable the editor to seek further advice.

(2) The reviewer can, on receiving a manuscript for the second time, return it immediately with the explanation that he has previously reviewed the paper. Whatever the actual circumstances may have been, the editor of the second journal is apt to assume that the manuscript has been rejected by another publication. In this case, the author is again at a disadvantage,

especially if he has revised the manuscript so as to respond to previous sound criticisms.

(3) The reviewer can return the manuscript immediately without other explanation than that he is unable to review the manuscript. Such a statement will not necessarily prejudice the opinion of the editor of the second journal.

Exercise of option (2) or (3) may, under certain circumstances, handicap the second editor. The reviewer may be, for example, one of the few experts available in a given field, or he may, as mentioned previously, be denying to the second editor what, in fact, is an extremely sound and perceptive evaluation.

Thus, no ideal solution exists for the dilemma posed when a reviewer who has already submitted a negative evaluation of a manuscript to one journal is asked to review the paper for a second publication.

Reviewing

CHAPTER 6

Technical and administrative aspects of manuscript review

Certain technical and administrative procedures will facilitate effective review of manuscripts submitted for publication. The experienced editor will usually adopt those technical and administrative procedures that are most suitable for a particular journal. Regardless of variations, however, he will need to consider each of the functions in the following procedures.

FILING SYSTEM

Four different types of files are recommended, each with a separate function.

Master file

The master file, which is best kept either as a card index or as a notebook, contains all information pertinent to the processing of a manuscript from the time of its receipt in the editorial office to its final disposition (publication or rejection). The file should be arranged sequentially by manuscript number—assigned by date of receipt of the manuscript. Following the number should be the names of the authors, their addresses, the title of the manuscript, the name and address of the person to whom editorial correspondence is to be directed, the date of receipt, and a description of the article (number of pages of text, number of references, number of tables, number of figures). A check list should also be included for various features of a manuscript that may be required for publication: abstract or summary, written permission to reproduce published material or cite unpublished data, identification of key words for index. (See page 78.)

Next, information should be recorded concerning the progress of the manuscript: date of receipt; names of reviewers to whom the manuscript was sent; dates of mailing of the manuscript to, and of its return by, the reviewers; date on which the editorial decision regarding acceptability or need for revision was forwarded to the author; date of receipt of revised manuscript; names of reviewers and dates pertaining to evaluation of the revised manuscript; date of final acceptance and forwarding to the publisher.

Manuscript file

Each manuscript, and copies of all correspondence relating to it, should be kept in individual folders in an appropriate file. Folders should be arranged sequentially by manuscript numbers. It may be useful to subdivide the manuscripts in either or both of two ways. First, they may be subdivided by their status in the reviewing process: (1) outstanding manuscripts, that is, those being reviewed by reviewers; (2) manuscripts awaiting revision by the authors; (3) accepted manuscripts awaiting publication; and (4) rejected manuscripts.

A second way of categorizing manuscripts is by type of article: original contributions, case reports, editorials, letters to the editor, reviews. Alternatively, editors of journals with a field-editor system may categorize manuscripts by discipline: cardiovascular field, endocrinology, epidemiology.

Author file

A separate card file, containing the name and address of each author and the number of his manuscript, should be filed alphabetically. This file facilitates communication with authors who inquire about their manuscripts without mentioning the titles. It also obviates confusion in communication with authors who have more than one manuscript under consideration simultaneously. Noting the names of the reviewers on the author file card is helpful.

Reviewer (referee) file

Some editors will want to have two separate files for reviewers, one classified alphabetically by name and the other by specialty, whereas others may require only an alphabetic listing. In addition to the name and address of the reviewer, each card should contain the assigned number of the manuscript he has been sent and the date on which it was sent and returned. Like the author file, this file will facilitate communication, and will minimize the chances of burdening one reviewer with excessive manuscripts at a particular time. Recording on the reviewer's card an evaluation of his performance on each manuscript provides guidance to the editor in assigning future manuscripts. Finally, this file will facilitate periodic publication of the names of reviewers, if such publication is a policy of the journal.

FLOW CHART FOR A SUBMITTED MANUSCRIPT

Before reading this section, the reader may wish to examine a sample flow chart (see page 112) showing the progress of a manuscript from the time it arrives in the journal editor's office until it is published or rejected.

When a new manuscript is received in the office of the editor, it should first be assigned a number. This number is recorded on the various cards already described and on each piece of paper associated with the manuscript. This number must be written on the manuscript, but it is a courtesy to the author to record it in pencil, since the author may wish to remove that number if his manuscript is rejected and he decides to submit it elsewhere.

After the number has been assigned, the filing systems mentioned earlier should be initiated. A letter or postal card should be sent to the author, acknowledging receipt of the manuscript and advising him that he will be informed of a decision after review. Promptly upon receipt of a manuscript, the editor should examine it with several purposes in mind. First, he can determine whether it is clearly inappropriate for the journal, in which case he can return it promptly to the author with an explanation of why it is inappropriate (such as out of scope). Some editors may suggest another journal for which the manuscript might be more suitable. Promptness is essential for this first step; understandably, some authors may react angrily when, after waiting a month or more, they learn of a decision that could have been made immediately.

In his initial review, the editor should also examine the manuscript to see if it is complete. If it is not, and if he is interested in the manuscript, he should promptly ask the author to supply any deficiencies that would prevent adequate review, such as missing pages, illustrations, or references. These omissions should, of course, be supplied before the manuscript is sent to reviewers. Other omissions, such as abstract, summary, or written permission to cite unpublished data, may not prevent adequate review and can be rectified during the reviewing process.

Next, the manuscript should be sent out for review (see "The Manuscript Reviewing System" and "Guidelines for Reviewers"). A mechanism should be established to remind the editor to communicate with reviewers who have not returned manuscripts after a reasonable (usually specified) interval.

It is the policy in some journal offices to have most papers reviewed by an in-house staff rather than by outside reviewers. The types of communication with reviewers are, however, generally the same. The discussion that follows assumes outside review.

When the reviews are received, the editor must decide on the acceptability of the manuscript. If the two initial reviews are largely in agreement, the editor will ordinarily, although not obligatorily, proceed in accord with the advice of the reviewers. If the two reviews show major disagreement, the editor may either arbitrate, himself, or seek a third review. The editor should keep in mind that the reviewers are not always right in their judgments. The decision regarding acceptability will usually place the manuscript in one of four categories: (1) acceptable, (2) acceptable with suggestions for revision, (3) acceptable contingent upon specific mandatory revisions, and (4) not acceptable.

The editor should then communicate his decision to the author. If the paper is acceptable as submitted, the editor need not return the manuscript, although he may wish to advise authors that minor changes may be made to make the manuscript conform to the journal's editorial style. All three other decisions require return of the manuscript to the author. If the paper is not acceptable, most editors will inform the author of the specific grounds for rejection, although some prefer not to do so, to avoid the arguments, rebuttals, irate telephone calls, and appeals for reconsideration that occasionally ensue. We have discussed a related problem in considering whether the reviews should be sent to the author. If the paper is acceptable subject to optional or mandatory revision, the editor should state precisely what is expected of the author. If the requested revision is simple and straightforward, the editor may agree to process the paper for publication once the changes are made. If, on the other hand, the recommendations are for major revision, requiring, for example, further experimentation, the editor should advise the author that the revised manuscript will have to be re-evaluated by reviewers. The author should also be asked to state, in a covering letter, how his revision accommodated the recommendations by citing pages on which he made changes, and to specify any editorial recommendations that the author rejects, indicating his reasons for doing so.

When the revised manuscript is received, the editor must again decide on its acceptability. At this point, the editor may choose to make the decision himself or to request consultation by one or more reviewers (see "Duplicate Reviewing"). Occasionally, it is necessary to return the manuscript to the author two or more times for further modification.

Once a paper has been accepted for publication, the author should be notified and the manuscript turned over to the redactional staff (see "Copy-editing"). After redaction has been completed, the editor should review the paper again to determine whether further corrections are needed. It is then sent to the printer, who sets it in type and produces galley and page proofs.

Most journals send galley or page proofs to the author for examination and correction. Any points that seem unclear to the redactional staff should be noted on the proof as queries to authors. Some editors send the edited typescript to the author with galley proofs. In such instances, the editor may wish to request the return of both, to determine how many additional changes the author has made beyond those necessitated by the editorial staff or printer. The author should be advised that he is responsible for proofreading and, thus, for any errors in the proof that he does not detect. The editor should have an opportunity to approve the proof after the author has made his corrections.

Artwork, roentgenograms, electrical tracings, photomicrographs, and other such illustrations are generally returned to the author when no longer needed by the editor and publisher.

CHAPTER 7

Ethical experimentation and the editor

Human experimentation

Investigators conducting biomedical, behavioral, and social studies on human subjects are expected to observe certain ethical standards designed to safeguard the rights and welfare of the subjects. Since this principle is widely accepted by the scientific community, as well as by society at large, all organizations whose functions impinge on such research should, by word and deed, support adherence to ethical standards. When editors evaluate the acceptability of the reports of such research for publication, they should, therefore, apply ethical as well as other criteria. To meet his responsibilities, the editor will require not only a mechanism for determining whether ethical standards have been maintained, but also a policy for disposition of manuscripts that do not meet these standards.

MECHANISM

In the United States, virtually all major institutions engaged in research on human subjects have institutional review boards (IRBs) to evaluate and certify the ethical character of experimentation. Such review not only serves as a safeguard against the misuse of human beings in research, but also helps protect investigators against charges of abuse. For research conducted or supported by the Department of Health, Education, and Welfare (DHEW), the requirement for such boards is described as an amendment to Title 45 (Public Welfare) of the Code of Federal Regulations. It is presented as Subtitle A, Part 46, and published in the *Federal Register* 40(50):11854-11858, March 13, 1975. Although these IRBs are required by law to review only research conducted or supported by DHEW, they do, in fact, review virtually all research involving human beings that is done in major institutions regardless of the source of support. At the time of this writing, Federal regulations are being revised to require that all research supported by DHEW—*including that conducted outside the United States*—be reviewed

29

by such IRBs and meet the ethical standards prescribed in DHEW regulations.

Thus, for most research reports submitted for publication, the editor may first wish to know if the research was approved by an IRB. Some editors may choose to accept that approval and look no further, whereas others may feel that such approval does not necessarily ensure that the research described meets the ethical standards of the journal. The actual research being reported may, in fact, differ substantially from that approved by the IRB, as, for example, when preliminary findings or circumstances during early phases of the research cause the investigator to change his experimental design without further consultation with the IRB. Moreover, an occasional unscrupulous investigator may deliberately deceive the IRB by describing one protocol while conducting his research according to another. Finally, some IRBs may fail to maintain standards high enough to suit the editorial boards of some journals. Accordingly, the editor may ask members of the editorial board and other manuscript reviewers to comment specifically on whether the described research raises any question about ethical propriety.

When research is conducted without IRB approval, the editor will have to use some other mechanism to ensure ethical propriety. In such cases the author may be asked to state the criteria as well as the peer review mechanism (if any) used in approval of the ethical aspects of the work.

Recommended procedures

(1) The published report of any research in which human beings have been used should contain a footnote indicating the mechanism used for reviewing the ethics of the research conducted. Ordinarily, this will signify endorsement by the IRB. [If the research was not reviewed by an IRB, the editor may use Recommended procedure (2).] The footnote should indicate that the research was approved by the IRB of (the institution) on a specific date, that is, the last date on which the IRB reviewed the project. The editor will thus know what ethical standards were in effect at the time. The date is important because standards, as expressed through regulations and guidelines, are evolving rapidly and being published frequently.

The author should be asked to provide, with his submitted manuscript, suitable documentation of this approval, such as a photocopy of the IRB's statement of approval. Both of these requirements (footnote and documentation of institutional approval) should be specified in the published Information for Authors. The purpose of the documentation is the same as that required for permission to reproduce copyrighted material or publish personal communications; it does not imply mistrust of the author.

(2) If the research has been carried out without IRB approval, the author should be asked to state how the ethical aspects of the investigation were evaluated, so that the editor may have a basis for an editorial judg-

ment. In some instances, the author's review group may differ from that required by DHEW, or the author may have had no review at all. If the peer group or the author himself has adopted a well-known code, such as the Declaration of Helsinki, or if DHEW regulations were observed without the filing of a general or special statement of assurance, the author should so state. In such a case, the editor may wish to follow Recommended procedure (1). Alternatively, the author may be asked to explain in some detail the ethical reasoning supporting his decision to conduct the research as he did.

(3) If, despite such statements of IRB or other appropriate approval, the research seems to the editor or reviewers not to have been conducted in accordance with ethical principles, the editor should inform the author that a question of ethical propriety has arisen, and should request a copy of the original proposal of research as it was approved, along with a copy of the ethical guidelines used. Thus, in the case of an IRB operating according to DHEW regulations, a request may be made for a copy of the general or special statement of assurance. In other cases, the guidelines may be requested as described in Recommended procedure (2).

(4) If, after having received the information specified in Recommended procedure (3), the editor or his advisers remain in doubt about the propriety of the research as *actually carried out,* the editor should ask the author to have the present chairman of the responsible IRB or other reviewing committee submit a signed statement certifying that he has personally reviewed the actual research and has found it to be ethically proper.

These procedures will usually enable the editor to decide whether the research described in the article has been conducted ethically. Whatever the outcome of the procedures, the decision regarding acceptability remains the editor's responsibility.

ETHICALLY QUESTIONABLE MANUSCRIPTS

The most difficult problem for the editor is to decide about disposition of a report that contains the suggestion or clear evidence of an ethical impropriety, but that, in all other respects, meets the criteria for publication. By the time an article is submitted, the research described has already been done, and the editor's primary responsibility at this point is to act most effectively to minimize the frequency of future unethical research. A uniform policy that is applicable to all articles is difficult to formulate; each must be considered individually. In making his decision, the editor should consider the following points. Publication of an article without editorial comment may be taken to imply that the editor and his reviewers condone the unethical experimentation. Refusal to publish, on the other hand, may have adverse consequences: it may deprive the scientific community, either per-

manently or temporarily, of valuable knowledge. Research conducted un-
ethically can generally be repeated ethically, but is it ethical to expose a
new group of subjects to risk to obtain information that is already available?

Ethical standards are not absolute, and many concern matters on which
responsible and thoughtful people disagree. Nor are ethical concepts station-
ary; what was previously unacceptable may now be condoned, and vice
versa. Editorial policies should permit expression of legitimate differences of
opinion. They should also allow, or at least not obstruct, disclosure of
controversy and commentary that will not only improve our understanding
of currently acceptable ethical standards but also contribute to their con-
tinuing evolution.

Two major issues that must be considered in the ethical assessment of
research are (1) risk-benefit relations and (2) informed consent. The editor
is in a strategic position to make an independent judgment on risk-benefit
relations, that is, whether the risk to the individual subject is warranted by
the potential benefit, preferably to the subject, but occasionally to the general community. On the other hand, the editor is in a poor position to
determine whether truly informed consent was obtained under conditions
in which the subjects were free from coercive influences. Thus, the editor,
having based a favorable judgment on a presumably acceptable risk-benefit
ratio, may learn, after the paper has been published, that the procedures
were conducted without acceptable informed consent. The editor is then
in the position of seeming to have condoned unethical research. Conversely,
rejection of a paper based on an apparently unfavorable risk-benefit ratio,
but without full knowledge of the consent negotiations, may suppress the
reporting of ethical research. A strong, although not unanimous, opinion
affirms the right of the fully informed, uncoerced person to subject himself
to research in which he takes substantial risks but which may yield no direct
benefit to him.

Authors assume that their communications with editors about the acceptability of submitted manuscripts are confidential. Accordingly, the
editor should inform the author before he communicates with the author's
institutional committee or anyone else about the ethical aspects of the research [see item (4) under Recommended procedures]. At this point, the
author retains the option of withdrawing his paper from further consideration. Similarly, any decision to reject a manuscript on ethical grounds should
not be communicated by the editor to any third party.

If the author should choose to withdraw his paper from further consideration, the editor is faced with a serious ethical dilemma. Is it the responsibility of the editor to expose unethical research activities when the
author, being unwilling to have his article published along with a critical
editorial, withdraws it from further consideration or when the editor rejects
the manuscript in which the results are reported? Some editors may feel

that they have satisfactorily disposed of their obligations by denying the article publication in their journals. Other editors may feel that their obligation goes further—either that they should publish an editorial criticizing the research even though they have declined to publish a report of that research or that they should inform the author that if they see the article published elsewhere, they will write such a critical editorial about it. Since the editor may, at some point, be faced with this dilemma, he should formulate a plan to deal with it and should publish that plan as a part of the policy on ethical requirements for publication in his journal (in the Information for Authors). An editor who criticizes an article that he rejects and that he thinks may be published elsewhere may be liable to criticism or even litigation if he fails to inform prospective authors of his plans.

When an editor questions the ethics of an investigator, he may damage the reputation of that investigator. Under some circumstances, the editor may even be called to account for ill-advised actions as a respondent to a suit for libel. If, therefore, an editor wishes to comment publicly on the ethical aspects of an article, he should avoid branding the investigator, and should comment only on a specific issue raised by the article. Generally, it is preferable to state the problem as a question rather than as an indictment. An editorial in which the ethics of a specific maneuver is questioned, for example, might begin as follows: "Because we wondered about the advisability of (mention the specific aspect of the research project that is in question), we have invited the authors, whose comments appear on page ___, to explain why they consider this procedure appropriate. Other interested readers are also invited to comment." Less desirable would be an indictment such as: "We condemn the following procedure because. . . ."

Recommended policies

For an article that the editor considers scientifically sound but ethically questionable:

(1) The editor may propose to the author that the article be published with an accompanying editorial in which the ethical questions are raised. In this case, the author should be invited to prepare a rebuttal for simultaneous publication.

If the author agrees to have a manuscript published with an accompanying editorial in which the ethical questions are raised with or without his rebuttal, additional questions arise. Specifically, should the reprints of the original article be prepared as a unit with the critical editorial and any rebuttal that may have been published? The editor has the option of requiring that this procedure be followed, but he should include this requirement in his published policy on ethical criteria. Such a policy would lead to a variety of consequences, some of which may not be anticipated. For example, it would break up the page format of the reprint (unless it were planned to publish this material in sequence in the journal). It would also add to the cost of the reprints, and some authors may object to paying for reprinting of the editor's comments and the rebuttal. Alternatively, the editor may offer the author the option of having the critical editorial and rebuttal made part of the reprint, with negotiation of appropriate financial arrangements. Finally, the editor may allow the author to have his article reprinted without the editorial(s). Under the third alternative, according to the editorial policy of the journal, an editor's note may or may not be added, to indicate that an ethical question had been raised in relation to this research, with exact page references to the critical editorial and any rebuttal.

(2) The editor may decide that it would not be in the best interests of the journal or of the scientific community to publish the article with or without commentary on the ethical issues. In this event, he should inform the author that the article has been rejected on ethical grounds.

If, in the course of considering the first alternative, the author chooses to withdraw, the net effect will be the same as that of the second alternative. The first alternative will thus be followed to completion only in those situations in which both the editor and the author feel they have a good case. These are exactly the cases in which the publication of the editorial and rebuttal are likely to yield the greatest benefits to the scientific community.

PUBLICATION OF ETHICAL REQUIREMENTS

So as to reduce the liability to legal suit, the editor should publish his policies and procedures for assessing ethical aspects of research involving human subjects, and reference should be made to the published account in the Information for Authors. Discussion should be sufficient to indicate the anticipated imperfections of these policies and procedures. Such published policies will avoid the appearance of condoning unethical procedures. A detailed account of the procedures followed by the editor in forming his judgment should put investigators on notice that the ethical aspects of their work will be carefully scrutinized and that they should cooperate with their institutional review boards.

In the publication of the journal's policies and procedures, all potentialities for editorial action adopted by the journal should be made explicit. This should help avoid surprises that may cause authors to be angry or litigious, or both.

Animal experimentation

It is recommended that a journal's Information for Authors specify that an author who submits a manuscript reporting work in which experimental animals were used should:

(1) state in a covering letter that he has adhered to provisions of Public Law 89-544, as amended by the Animal Welfare Act of 1970 (Public Law 91-579); DHEW Publication No. (NIH) 74-23, *Guide for the Care and Use of Laboratory Animals,* revised 1972, second printing 1974; DHEW Animal Welfare policy HEW TN 73.2 (May 14, 1973), if funded by DHEW; and related animal welfare rules and regulations henceforth issued by Federal, state, or local agencies and one's own institution.

(2) in the text of the paper, describe methods and procedures to reflect the degree of adherence to ethical principles. In particular, the use of anesthetics, analgesics, and tranquilizers to avoid all unnecessary pain and distress should be described fully. When pain or distress occurs during an experimental procedure in which anesthetics, analgesics, and tranquilizers cannot be used without interfering with the nature of the study, full justification should be made of the deviation from accepted practice. Use of drugs or other procedures for killing of animal(s) at the end of an experiment should also be described.

The editor should include in the Information for Authors any other special ethical requirements of his journal for reports of animal experimentation (see "Information for Authors").

CHAPTER 8

Multiple publication

Editors of many journals in the life sciences prefer not to accept material whose substance has already been published, has been accepted, or is being considered for publication elsewhere. (Review articles are an exception to this scientific editorial principle.) The multiplicity of forms in which biologic, and especially biomedical, material may be printed, however, makes a categorical definition of "already published" difficult. In addition, the circumstances of any apparent prior publication must be taken into account, especially since authors do not always have complete control over the publication of their data or concepts. Despite these complexities, the following General Guidelines may be useful to editors in deciding whether or not the contents of a manuscript should be classified as "already published."

GENERAL GUIDELINES

(1) Classification of an article as "already published" is warranted if it has appeared in a general or specialty journal, whether basic, applied, or clinical.

(2) Classification of submitted material as "already published" is not warranted by previous publication of:

(a) an abstract of a paper presented, or offered for presentation, at a formal meeting, or

(b) a report issued by requirement of a governmental agency or comparable institution.

(3) Classification of material requires individual consideration when it has appeared in other forms of publication such as monographs, editorials, letters to the editor in scientific publications, dissertations, preliminary notes, or such unreviewed publications as widely distributed proceedings of societies, meetings, and conferences, and reports in scientific and medical news media. The following criteria may be helpful to editors in classifying such material as "already published":

(a) the manuscript adds no basic concept or important new information (including tables and figures) to material that has appeared in one of the forms or media listed;

(b) the circumstances suggest that the prior publication in a news medium was actively promoted by the author(s).

DISCUSSION

These guidelines are designed to avoid the following disadvantages of multiple publication: (1) removal of the quality of newness from a scientific journal by previous publication elsewhere of the essence of the material it contains, (2) cluttering of the scientific archives with rehashes, (3) premature publication in the scientific or lay news media of research findings that may not be scientifically valid and may therefore mislead or be harmful to society, and (4) copyright liability. These guidelines may assist editors of scientific journals who are faced with difficult decisions about whether a certain manuscript meets a journal's criterion of newness of information. They do not purport to provide definitive rules as a basis for such decisions, since we recognize that an editor must judge each case individually and make the decision he considers most appropriate for his journal.

The editor may wish to inform potential contributors of the position of his journal on multiple publication by including a statement in the Information for Authors and requesting that authors advise the editor, in their letters of submission, that the material has not been published elsewhere. For material that has been published or is contemplated to be published in abstract form, the editor may wish to request that a copy of the abstract accompany the full manuscript, so that he can make a suitable judgment. If any form of preliminary publication other than an abstract or condensation of not more than 400 words has appeared anywhere or is contemplated, a reprint or photocopy of the actual or proposed publication should accompany the manuscript being submitted.

For the journal that publishes reports of original research exclusively, multiple publication may be only an occasional problem. For the general journal or the clinical journal, one of whose functions may be educational, the issue may be more complex. We recognize the legitimacy of reprinting outstanding articles or reproducing visual components of articles when this is done by agreement between the authors and editors involved (and with the permission of the publisher) and when a valid reason for such reproduction exists. It is the abuse of republication that we believe should be discouraged. Approval should always be obtained to use excerpts from an article or to delete any portions of it before reproduction; these circumstances should be clearly indicated in the reproduction. Reprinted material should always be clearly labeled as such, and this information should not be suppressed in the bibliographic reference.

Simultaneous submission of an article to two or more editors is to be discouraged; it multiplies not only the expense to journals of processing a manuscript, but also the time of reviewers in examining and evaluating it for publication. Since most reviewers serve journals without remuneration, this duplication of effort is an unnecessary imposition on their services. Moreover, if the manuscript should be accepted by two or more journals, additional difficulties would arise in copyright privileges and in scheduling of publication.

Although it is difficult to prevent anyone in attendance at a public meeting from tape-recording a talk and photographing the speaker's slides, reporting such a talk verbatim in the news media or elsewhere, or publishing or duplicating the slides without the speaker's permission, is considered improper because it represents a breach of common law property right. If this is done without the knowledge of, or encouragement from, the speaker, it would be unfortunate if it jeopardized the speaker's chances of publishing his manuscript in a peer-reviewed scientific journal. All concerned should, nevertheless, be aware of the implications of the copyright held by the news medium or other previous publisher involved.

We recognize the "public's right to know," but believe that the phrase subsumes a responsibility on the part of scientists, physicians, editors, publishers, and science writers to make certain that information disseminated through the scientific and lay news media is accurate and valid. Journalists should guard against premature reporting of information that may not be scientifically valid and that may therefore mislead or be harmful to society. The "public's right to know" is not therefore necessarily synonymous with "instant reporting," and a journalist often better serves his readers by delaying publication until the substance of the report has passed the critical scientific analysis of peer review.

The traditional mechanism for peer review is the referee system used by journal editors before publication of manuscripts, a system that has served science well. Whereas most professional organizations have program committees that select the papers for inclusion in programs for meetings, the full manuscript is not generally available to the committee, and the information in the abstract that the author provides is often anticipatory, scanty, or considerably different from his oral presentation. For this reason, the function of the program committee cannot be considered equivalent to peer review.

If professional societies did not require, or request, that their program speakers submit full-length manuscripts in advance for use by the press room, much of the problem regarding prior publication in the news media would be solved. When a report appears in the news media, even in abbreviated form, readers of scientific journals in which the full manuscript may later appear are likely to consider that they are already familiar with the material and may therefore miss important points of information that were omitted in the popular version.

The science writer, his editor, and his publisher have an equal obligation with that of the author and journal editor to exercise the highest ethical and intellectual responsibility in the dissemination of scientific information. By fulfilling this obligation, journalists can protect society from the harmful effects of falsely announced "breakthroughs," unsound scientific concepts, and other misinformation.

CHAPTER 9

By-lines and acknowledgments

BY-LINES

We consider two recurring problems:

(1) *Should editors try to regulate the choice of authors' names that should appear at the head of an article?*

There is a tendency for naive authors of a paper to include, as a sort of compliment or reward, the names of "co-authors" who made only peripheral contributions to the work described—for example, obtaining the research grant that supported it, discussing the interpretation of the results one afternoon, or providing technical assistance at an unusually high level. One disadvantage is the proliferation of papers bearing a disproportionately large number of names, with consequent cumbersomeness of citation and increased possibility of error. Other authors with less pure motives may produce inappropriate by-lines either by omitting the name of a substantial contributor or by adding an illustrious name in the hope of attracting more favorable notice for their work.

Because authors may, without giving the matter much thought, bring undesirable practices into the composition of by-lines, we suggest giving guidance such as the following in the Information for Authors:

"The by-line should contain the names of those who have contributed materially to the work and its report; those who have participated only in an advisory or supporting capacity should be thanked in the Acknowledgments."

Some authors wish to reward unusually gifted or devoted technicians by elevating them from the Acknowledgments to the by-line, but do not quite want to give them co-author status. They solve the dilemma by adding, after the names of authors, "With the technical assistance of. . . ." Some journals permit or encourage this arrangement, but thereby make the situation bibliographically messy; others who cite the work find it difficult to decide whether to list these names with those of the first-given authors or not. We do not recommend this method of rewarding technicians for devoted service.

40

(2) *Can a standard order of authors' names be prescribed?* Some editors have proposed requiring authors to be listed in descending order of magnitude of contribution, or with the most active author first and second most active last, or even alphabetically. In our opinion, the editor must let the authors decide for themselves how the order of names is most probably going to be interpreted, and construct the by-line accordingly.

ACKNOWLEDGMENTS

The author's choice of whom he will thank and where he will stop has traditionally been his prerogative. The editor may not wish to interfere with this prerogative, although some style manuals discourage acknowledgment of technicians and other remunerated personnel for carrying out routine operations, or resident physicians who merely care for experimental patients as part of their hospital duties. Editors may wish to discourage such acknowledgment, but will probably not make a big issue of it.

On the other hand, acknowledgment of intellectual or professional participation short of co-authorship does require an editor's guidance. We recommend that a statement such as the following appear in the Information for Authors: "The author submitting an article should state in a covering letter that the content and wording of acknowledgment of help (for example, the supply of rare materials) or criticism have been approved by those whose help is acknowledged." We regard this precaution as essential because the thanking of a colleague for his help may be interpreted to mean that he approves of the article. If he does not approve of it, he may object to being so thanked.

CHAPTER 10

Publication of dates for manuscripts as an index of priority

The date of a scientific paper containing any pretensions to discovery is frequently a matter of serious importance, and it is a great misfortune that there are many most valuable communications, essential to the history and progress of science, with respect to which this point cannot now be ascertained. This arises from the circumstance of the papers having no dates attached to them individually, and of the journals in which they appear having such as are inaccurate, i.e. dates of a period earlier than that of publication. . . . These circumstances have induced me to affix a date at the top of every other page, and I have thought myself justified in using that placed by the Secretary of the Royal Society on each paper as it was received. An author has no right, perhaps, to claim an earlier one, unless it has received confirmation by some public act or officer. (Faraday, Michael: *Experimental Researches in Electricity*. Vol. I. London, Taylor, 1839, p. iv)

Because novelty is highly valued in science, a scientist's reputation rests in part on his ability to formulate new ideas, methods, or proofs. Priority for new concepts, new techniques, or other new knowledge is therefore a major incentive in scientific research. With disclosure of his new concepts, however, the scientist relinquishes certain prerogatives. When he submits his manuscript for publication, he exposes it to the scrutiny of the editor and usually to that of unknown reviewers, and he surrenders the right to submit it for publication in another journal. In addition, publication in a scientific journal may require several months to a year or more after the article is received in the journal's editorial office, while it is being reviewed, copyedited, and printed. In view of these conditions of publication, the date of receipt of a report of new knowledge in an editor's office for dissemination to a wide scientific audience is important as an index of priority.

The editor, as guardian of the author's intellectual property, should be aware of the question of priority. There are three interested parties: (1) the author who wishes to protect his priority, (2) the editor who wishes to protect the reputation of his journal against imputations of procrastination in editorial handling of manuscripts or of lack of integrity on the part of his reviewers, and (3) other scientists whose claim to priority may be undermined by assigning a debatable date to an article.

The problem of selecting a date to be attached to a published article, however, is not as simple as it might first seem. The following chronology of events involved in a scientific project from conception of the idea through the research endeavor to the publication of the final report illustrates the complexity of the decision. The numerous dates listed emphasize the importance of keeping accurate records in the editorial office of all dates associated with the processing of a manuscript for publication and of indicating in the journal's Information for Authors the editorial policy concerning the dating of articles.

(1) DATES BEFORE SUBMISSION OF THE MANUSCRIPT

Dates before the submission of a manuscript are rarely considered by editors, but they may be important for legal reasons, as, for example, when patents are involved. Such dates are often recorded and witnessed in laboratory notebooks; this practice is more frequent in industrial research than in academic research.

(a) Date of original idea

The date of the original idea is the date when the idea is born in the author's mind. He may mention a new concept at a meeting or in informal discussion with colleagues, and may begin working on the project immediately, or he may delay this work for some time. Priority established on this basis is not the responsibility of the editor.

(b) Date of completion of the first definitive experiment or of completion of the work

The dates recorded in the author's notebooks are sometimes used to verify the time of completion of an experiment or research project. Some authors date illustrations or insert the date of experiments in the manuscript to ensure priority, but some editors refuse to publish such dated illustrations or texts.

(2) DATE OF CIRCULATION OF PREPRINT

Some authors send copies of the typescript of an unpublished article to friends and colleagues, either to obtain criticism or to make the information in the article available in advance of publication. This practice may give rise to difficulties if the preprint prompts other authors to rush into print. The practice may also be used as a means of claiming priority. Neither of these results is the responsibility of the journal editor.

(3) DATE OF SUBMISSION OF MANUSCRIPT

(a) Date on letter of submission

The date on the author's letter of submission, with which the manuscript is enclosed, ostensibly indicates when the author released his information for circulation to the scientific community. In questions of priority, however, it would be difficult to prove that the letter had not been pre-dated.

(b) Date of postmark

Because a postage meter may be set to print a date earlier than the actual date of posting, the date shown on metered postage is not necessarily authentic. A postmark applied by the post office may, unfortunately, show a date some days later than the date on which a manuscript was mailed.

(4) INITIAL DATE OF RECEIPT OF MANUSCRIPT

The date on which the manuscript is received in the editorial office is important, since it is the date on which the editor's responsibility begins. Unfortunately, delays in the mails or in institutional mail rooms may mean that the date of receipt of an article by the editor may be many days after the article has left the hands of the author. Nevertheless, the date on which the article actually arrives in the editorial office is a practical and widely used date. If this date is used, it should be the actual date of arrival of the article in the editorial office, not a possibly later date when the editor's secretary finds time to acknowledge the article or a possibly still later date when the editor sees the article for the first time.

(5) DATES OF EDITORIAL ACTION

(a) Firm acceptance of manuscript

Some time will elapse between receipt of an article and its acceptance by the editor, even if the article is accepted exactly as submitted or subject to only minor, nonsubstantive typographic or stylistic revision. Such delays may arise, for example, from postal delays during circulation of the manuscript to reviewers or from procrastination on the part of the editor or reviewers. It is, therefore, probably unfair to print the date of acceptance without the date of receipt as well, especially if little or no revision has been required. Publication of both dates may also protect the editor from charges of unwarranted delays. The editor will, of course, want to keep accurate records of the dates on which manuscripts leave and return to his office during the reviewing process.

(b) Provisional judgment

Relatively few articles are accepted without minor or major revision. Some journals invariably publish two dates, that of receipt of the original version and that of receipt of the revised version. Yet a long interval may elapse before receipt of truly inconsequential revisions as a result of postal delays, procrastination by reviewers or the editor, or other reasons over which the author has no control. To publish both dates in such a situation may unfairly suggest that the author had been required to make drastic and time-consuming changes before his article became acceptable. Certainly, omission of the date of receipt of the original version in such a situation would be unfair.

At the other extreme, extensive revision or even the performance of new studies or experiments may be a condition of acceptance, in which case the inclusion of the date of receipt of the original version may give the author an undeserved priority.

The magnitude of the changes effected by revision may warrant suppression of the date of receipt of the original version if the revision represents essentially a new article; in such case the revision should have a new date of receipt. Such action, which must be initiated by the editor, can raise delicate problems. If the editor anticipates this problem with regard to a particular article, he may prefer to decline the original article instead of accepting it subject to revision. In that case, the editor would inform the author that although the article has been declined, he would be glad to receive a *new* article extensively revised along the lines suggested by the reviews. Such an article would automatically receive a new "date of receipt of manuscript."

Because most articles fall between the foregoing extremes, requiring revision that is neither trivial nor momentous, many journals print the dates of receipt of both the original and the revised version(s) (including the second and third revisions if need be). The foregoing discussion shows, however, that printing these dates does not automatically solve all the problems of dating articles.

(c) Essentially acceptable version

When the original or a revised version proves acceptable, subject to a few essentially trivial and nonsubstantive changes, such revisions are ignored in printing of dates. If, therefore, the date of receipt of the original version, the date of receipt of the revised version, or both are used, it is generally understood that changes made subsequent to either date were inconsequential. It is, however, the editor's responsibility to make certain that such changes are essentially trivial and nonsubstantive. For this reason, if a date intended to reflect the date of acceptance is used, there is much to be said for using the date defined in 5(d), the date of receipt of the article as actually published.

(d) Version as actually published

A well-defined date is the date of receipt of the manuscript in the form (apart from copy-editing) in which it appears in the journal, that is, the date on which the editorial office receives from the author the version that is handed over to the copy editor to be prepared for the printer. If further changes of any kind are made by the author at the request of the editor, "the date of receipt of the version actually published" should be the date on which those final changes reach the editor. As with the date of receipt of the original version, this date should be the date of actual receipt, not the date on which the editor's secretary or the editor gets around to dealing with the manuscript.

Again, the interval between the date of receipt of the original version and the date of receipt of the version actually published may be long through no fault of the author, and yet such a long lapse may reflect unfairly on him. From the point of view of the editor, however, these dates are valuable, since they define exactly the period during which substantive changes for which he is responsible were made. Changes made thereafter by the copy editor should, generally, be confined to removal of grammatical improprieties and to application of journal conventions in matters of punctuation, capitalization, abbreviations, and the like (see "Copy-editing").

(6) DATE OF CHANGES MADE IN PROOF OR OF ADDENDA TO THE MANUSCRIPT

If the author makes substantive changes in the proof or if he appends addenda *after* final acceptance of the manuscript by the editor, this emendation should be noted in some way. Such changes are most easily handled if they appear separately at the end of the article either as a paragraph entitled "Addendum" that begins with the statement "Since the above article was submitted (or accepted for publication), . . ." or as a paragraph entitled "Note Added in Proof." In such instances, the addendum or note may bear its own date of receipt, but even if it does not, the reader and anyone interested in questions of priority will be alerted to the fact that the added material is of more recent origin than the rest of the article.

(7) DATE APPEARING ON THE COVER OF THE JOURNAL ISSUE IN WHICH THE MANUSCRIPT IS PUBLISHED

The putative date of publication of the journal, that is, the date printed on the cover, is important because it is the one used by most authors in citing the works of others. If the putative date of issue of a journal containing manuscript A is earlier than the putative date of issue of a journal containing manuscript B, the author of A will usually receive priority even if the date of submission of B preceded that of A, since most people cite only the date appearing on the cover of the issue of the journal, or the year corresponding to the volume in which the article appears. If, however, the dates of receipt of manuscripts are published, a careful writer will assign priority accordingly.

(8) DATE OF ACTUAL PUBLICATION OF THE ISSUE OF THE JOURNAL

The date that appears on the cover of a journal is not necessarily identical with the actual date on which that issue is published, that is, the date on which it is made available to the public. The date of actual release to the public can be of particular importance in patent disputes and in establishing priority in taxonomic descriptions (see "Indexes").

The *International Code of Botanical Nomenclature** states:

> Publication is effected, under this Code, only by distribution of printed matter (through sale, exchange, or gift) to the general public or at least to botanical institutions with libraries accessible to botanists generally. (Article 29)
> The date of effective publication is the date on which the printed matter became available as defined in Art. 29. In the absence of proof establishing some other date, the one appearing in the printed matter must be accepted as correct. (Article 30)
> The date on which the publisher or his agent delivers printed matter to one of the usual carriers for distribution to the public should be accepted as its date of publication. (Recommendation 30A)

The date of publication, that is, the date on which the journal becomes available to the public, may occasionally be different from the date that appears on the cover of the journal. In the first issue of each volume, some journals contain the actual date of delivery to the post office of each issue in the preceding volume. Even if this information is not printed in the journal, it is wise for the editors or publishers to obtain these dates and keep a record of them.

(9) DATE OF RECEIPT OF THE JOURNAL BY THE SUBSCRIBER

Persons concerned with priority should be aware that there may be a discrepancy between the date that appears on the issue of the journal and the date on which the issue is received by the subscriber. The issue may be delayed in being published, or strikes or postal delays may intervene; moreover, a subscriber whose copy arrives by surface transoceanic mail may receive it two to four months later than it is received by some subscribers. Thus the scientist who receives a preprint even before the article is submitted may read the article more than a year before the scientist who receives the journal by slow mail. During that year the second scientist may have had the same idea, conducted experiments, and written and submitted a similar article without any knowledge of the prior work. Such an author might be suspected of having advance knowledge of work he has never

---

*Utrecht, The Netherlands, A. Oosthoek Publishing Company, for the International Association for Plant Taxonomy, 1972, pp. 36-37.

heard of, and may be entirely innocent of any impropriety, even though the date of submission of his work is as much as three months later than the date of publication of the article containing the prior work.

GENERAL REMARKS

From the foregoing discussion, it should be evident that the problem of assigning a date to an article is not a simple one. As mentioned in an introductory statement, however, such dating is important to the author, the editor, and other scientists. In general, the date of receipt of the original version protects both the author and the editor; explicit identification of the date of receipt of a major revision or of addenda protects the editor and other scientists; and the date of receipt of the version actually published protects the author and the editor in the sense that it establishes a date after which neither the editor nor the author has made further changes in the article.

Disputes over priority cannot be wholly avoided by any system of dating of articles, but each editor should be aware that disputes can give rise to ill feelings or even legal problems. The editor should therefore establish, for his journal, the system that he believes will minimize such disputes and will protect all interested parties.

CHAPTER 11

Advertising

POLICIES

Some publishers of scientific professional journals use revenue from advertising to meet some or all costs of publication. Although advertising brings additional income, it also entails certain disadvantages: (1) even if the sponsoring organization has a tax-exempt rating, the income of the journal is subject to Federal taxes; (2) the unpredictability of advertising income (which is apt to fluctuate much more than subscription income) compromises the preparation of reasonably reliable budgets; (3) advertisers may try to influence the content of the journal by overt or veiled threats of canceling advertisements; and (4) the editor may be suspected of compliance with the wishes of advertisers even if his actual practice is beyond reproach.

A journal that accepts advertisements should have explicit policies governing their handling and publication. The editor should not only be well acquainted with these policies but should also know to what extent he is responsible for their establishment and implementation and to what extent these functions are in the hands of the sponsoring organization or its agent (executive editor or business manager). The editor also needs to be aware of the current postal regulations regarding advertising and should make sure his journal complies with them.

Content

The criteria for acceptable advertising should be defined with respect to (1) categories of items approved for advertising, and (2) substance of copy and illustrations submitted.

(1) Some journals accept only advertisements for professionally related items (laboratory equipment, drugs), whereas others accept advertisements for a wider range of categories but exclude certain products, such as alcohol and tobacco.

(2) The system used for assessment of the substance of promotional material should be specified. If the editor (as opposed to an agent of the sponsoring organization) has responsibility, he may review submitted ad-

49

vertisements himself or may delegate the responsibility to others. A favorite practice is to appoint an advertising committee charged with passing on the acceptability of advertising. Such committees, however, do not always operate efficiently because of possible bias, limited competence on the part of members who are specialists, and delays in committee decisions. Another system is to assign primary responsibility for evaluating advertising material to an (editorial) associate who would be expected to seek the advice of consultants as necessary. This associate should be a professional with broad qualifications appropriate to the interests of the publication.

The person or group responsible for evaluating the acceptability of advertising should be authorized to negotiate with the advertiser to effect changes required to meet the journal's criteria for advertising.

The content of pharmaceutical advertisements is influenced to a large degree by the regulatory activities of the Food and Drug Administration (FDA), but at present the FDA does not screen such advertisements. Unless this practice changes, publishers of journals are well advised to continue their own review of pharmaceutical advertising.

Amount

The total number of pages of advertising that will be accepted should be specified either in absolute terms or as a percentage of textual pages. Current postal regulations regarding this matter should be consulted.

Disclaimers

A disclaimer, stating that acceptance of advertising material does not imply endorsement, is usually included in information in the publication regarding editorial and publishing practices. Despite such a disclaimer, however, the tendency (not necessarily legal opinion) is to hold the publisher and editor responsible for the advertising content of a journal. Although the editor is not likely to be held legally responsible for improper advertising—except when the impropriety is serious and undeniable—he should exercise as much *de facto* responsibility as is available to him to ensure reliability of advertising.

PRACTICES
Placement

The placement of advertising should be specified: Is it to be stacked, that is, separated from textual pages, or is it to be interspersed in the text? If interspersion is decided upon, the degree needs to be specified. Interspersion of advertisements between articles has the advantage of allowing immediately succeeding articles to be torn out for filing without sacrifice of any portion of either article. On the other hand, many regard interspersion as functionally or aesthetically objectionable. Some journals include advertisements on the table-of-contents page, particularly if this is also the front cover, but such practice is decreasing.

Librarians prefer the stacked system because it permits them to have journals bound without advertisements. On the other hand, for historical purposes, a journal editor should make sure that his office retains at least one copy of every issue bound exactly as it is delivered to the subscriber (see "Binding Practices"). When some advertising is carried on the obverse of pages containing scientific material, those advertisements cannot, of course, be removed before binding, and the bound journal may then contain some, but not all, advertisements.

Pagination and index

Advertising pages should be numbered; a different system from that for the text, such as Roman numerals, is generally used. An index of advertisements and advertisers is also advantageous, but not absolutely necessary.

Special types of manuscripts

CHAPTER 12

Editorials

Editorials originally presented the viewpoints of the editor(s) or own-er(s) of a newspaper or journal, but over the years the scope of editorials has broadened. Thus an editorial may now offer an opinion or a proposal, per-haps even polemical in tone, or may dispassionately summarize current knowledge concerning a subject of interest to the readers of the publication. Many editorials use an intermediate approach: information is provided but is used, to a variable degree, as a basis for interpretation, extrapolation, evaluation, advice, or persuasion.

Although some publications restrict the subject matter of editorials to socioeconomic or legislative topics, or to the policies of the sponsoring organ-ization, others permit a wide choice of topics. A number of biomedical pub-lications, however, feature editorials on strictly scientific subjects that may or may not be related to an article appearing in the same issue.

Irrespective of the subject, a discussion in editorial form implies an au-thoritative source, often but not necessarily individual. The need for docu-mentation and for a standardized format in an editorial is less rigorous than in a conventional scientific report or review, but the need is greater for ele-gance and individuality of style. Although their length varies, editorials need not usually exceed 1,000 words, and rarely should require more than 2,000 words.

Because of the original purpose of editorials, the first person was ex-pressed in the plural (we, us), and the writer was not identified. These prac-tices are still widely observed but may be questioned today as artificial and anachronistic conventions. Many current editorials, particularly those that provide information about a scientific subject, do not represent the views of the editors or the publishers, but rather of an invited expert. For this reason, it would seem preferable for editorials to be signed, and to be written in the first person singular. Such identification also enhances the writer's sense of responsibility.

Similar considerations apply to editorials that express opinions. Even if a member of the editorial staff writes the editorial, his words do not neces-

52

sarily reflect the position or policy of the sponsoring organization or owner. Hence, to avoid misrepresentation, to help the reader identify the source, and to distinguish personal editorial opinion from official policy, all editorials in scientific publications are best identified as to author. If editorials are unsigned, however, the publisher may wish to print a routine disclaimer indicating that editorials do not necessarily reflect the views of the publication's owners or sponsors.

Although a catchy editorial title may attract attention, the title should, above all, be informative. Illustrations and tables are not commonly used in editorials, but the inclusion of such material is entirely appropriate.

Editorial titles should be listed in the journal's table of contents and in the index. If editorials are signed, authors' names should also be included in the table of contents and in the index (see also "Solicited and Remunerated Manuscripts").

CHAPTER 13

Abstracts, transactions, proceedings, minutes, news items

Abstracts that appear in scientific periodicals may serve different functions and therefore have different forms. Editors of journals that serve as official organs of publication for professional societies may be required to publish material pertinent to the activities of that society (Transactions, Proceedings, Minutes, News Items).

ABSTRACTS
Types

Abstracts fall into several categories. The designations given here represent some of the types used by different journals.

(1) *Abstract published at the beginning of the article and designed to acquaint the potential reader with the essence of the text.* It should be prepared with the same care as the remainder of the article, since the decision to read the article may be determined by the quality of the abstract. The abstract should be informative, rather than descriptive, that is, it should present, in brief form, the essential points made in the article. A mere *description* of the contents is uninformative and worthless: "Fifty cases of St. Louis encephalitis in the elderly are reviewed." For an experimental report, the abstract should include the purpose, subjects and techniques used, important results, conclusions, and their significance. It should contain nothing that is not in the article itself.

Authors are usually asked to prepare their own abstracts and are given guidelines as to length and format in the journal's Information for Authors. Abstract journals sometimes reproduce this type of abstract (see item 2). In such instances, the primary journal may ask the author to observe certain requirements of the abstracting journal as to length and style.

(2) *Abstracts of articles published elsewhere.* Some journals have a section devoted to abstracts of articles published in other periodicals, to allow the reader who does not have access to a wide variety of journals to keep

abreast of current publications. Other journals are devoted exclusively to the publication of abstracts of previously published articles. Because these abstracts stand apart from the original article, they should be completely self-contained, and sometimes, therefore, may include more information than those of type 1.

(3) *Abstract of paper proposed or accepted for oral presentation at a scientific meeting.* This type of abstract is a primary communication, since the material it contains has presumably never been published. Some organizations publish all abstracts submitted for oral presentation before a selection is made of those to be included in the program, whereas other organizations publish only those that have been evaluated by a committee and selected for inclusion in the program. In the first instance, abstracts may be committed to publication when received, whereas in the second, there may be a considerable delay from receipt of the abstract to its appearance in print.

Authors should be instructed in specific terms about the rules established by each organization for preparation of abstracts. Such rules may be published in the journal with the "Call for Abstracts," or authors may be instructed to write to the journal for the guidelines. When a form is provided, the rules are best printed at the top or on the reverse of the form itself. Organizations that distribute the "Call for Abstracts" to their members by mail should include the rules in the same mailing. (See pages 113-115.)

(4) *Abstract of paper already presented at a scientific meeting.* Publication of abstracts of papers presented at meetings allows those who could not attend to obtain the essence of the information presented.

Editorial responsibility

Editing of abstracts is a matter of independent decision for each editor, depending on constraints of time, space, and personnel. Whenever possible, the editor and staff should review abstracts with the same care as that used for a full-length manuscript. Regardless of type or purpose, the abstract should conform to the standards of quality, comprehensiveness, syntax, grammar, and language established by the editor. Abbreviations should be explained, and spatial limitations observed. When abstracts are submitted on special forms for photoreproduction, the editor must, of course, have the material retyped if he makes any editorial changes. When abstracts typed on such forms are to be photographed without change for publication, authors should be informed of this policy and should be advised that unless they read the abstract carefully, *themselves,* before submitting it, they may be embarrassed by errors in the published version.

The responsibility assumed by the editor for the content of the abstract varies considerably from one journal to another. In some journals, the editor reads every abstract to seek out errors in fact or typography, or other imperfections. The expanding volume of abstracts submitted for presentation at meetings has forced some editors to publish *all* such abstracts as submitted, provided they conform to the spatial limitations set by the journal, are reproducible photographically, and have comprehensible titles. Editors who publish large numbers of abstracts in this way risk being embarrassed by the possibility of an author's undetected insertion, in an abstract, of a political statement or a caustic comment about a rival researcher. Scanning titles will not, of course, disclose such inappropriate material, and the editor may therefore wish to publish a statement in the journal to the effect that the contents of abstracts are the sole responsibility of the authors.

Some journals have the abstracts set in type and published in the manner of full-length articles. This procedure is more expensive and time-consuming, and is likely to produce more typographic errors (a hazard each time a manuscript is published). Production deadlines usually do not allow for reading of proofs by authors.

Title

The title of the abstract deserves special attention. Some journals use a computer-developed permuted key-word-in-context index (see also "Indexes"). In an inclusion index, the computer is supplied with a large list of acceptable words, and any word in a title that is not on that list will be rejected (stop-word). In an exclusion index, the computer is supplied with a smaller list of unacceptable terms, which will be rejected if they appear in a title. The inclusion index is more expensive and perilous than the exclusion index, and must be updated more frequently. For either to be effective, the words in titles must be selected judiciously. The permuted index is not only much less costly than a hand-developed index, but is also rapidly developed and permits publication of the index in the same issue as the abstracts. It imposes a significant responsibility, however, upon author and editor.

Pagination

Many journals have a page-numbering system for abstracts that differs from the pagination of other editorial copy. This system facilitates early printing of abstracts, complete with indexing, well in advance of the publication of the program. One added advantage of the conspicuously different pagination is that those acquainted with the policy of the journal will recognize that any citation with such pagination is not a full-length article. Those unfamiliar with the system, on the other hand, may not realize that the reference is an abstract unless that word is included, in parentheses, in the citation.

With separate pagination, the publisher should be able to produce additional copies of the abstracts when the appropriate journal issue is printed. With some modification, these copies can serve as programs to be mailed to the membership of the society and to be distributed to other registrants at the time of the meeting. The saving in cost of printing makes this arrangement economical for societies with small memberships.

Cost

The expense for publication of abstracts can usually be charged to the society at publisher's cost. Details of this agreement should be worked out with the officers of the society and should be stipulated in a written agreement between the society and the editor or the editorial board. The editor's budget is thus spared the expense of publishing such material.

TRANSACTIONS, PROCEEDINGS, MINUTES, NEWS ITEMS

Other material that a society-owned or society-sponsored journal may be obliged to publish includes Transactions, Proceedings, Minutes, and News Items. Although responsibility for preparing this copy belongs to the officers of the society, the editor will want to review it just as he does everything else that appears in his journal.

Transactions represent a published record of actions taken and addresses delivered or papers read at an official meeting of the society. *Proceedings* may be synonymous with *Transactions* and may incorporate *Minutes* as well as actions taken by the society in the interval between official meetings. (Serially published proceedings or transactions are not included in this discussion because they are published independently of the journal proper.)

Occasionally, a society may wish to publish the discussions of participants in the Proceedings (or Transactions) of the journal. Such a policy adds several lengthy steps to the editorial process and may cause considerable delay in publication. Each discussion must be recorded, the recording transcribed, the transcript edited, and the edited copy retyped into publication copy for submission to the editor. The policy of some societies is to submit galley proofs or edited copy to the discussants for their approval. Although these procedures are the responsibility of the society and its officers, the editor must review each step of the process to ensure accurate publication.

The editor should assume responsibility for publishing accurate Proceedings and for arranging and meeting the production deadlines. Since Proceedings for a given society are usually in the same format each year, the editor should establish the details of the format with the copy editor, so that copy-editing can proceed efficiently for each meeting. For further details, the editor may wish to consult:

1. *Proceedings in Print* (periodical published by Proceedings in Print,
   P. O. Box 247, Mattapan, Massachusetts 02126)
2. *Medi-Kwoc Index* (periodical published by Washington University
   School of Medicine, Library, 4580 Scott Avenue, St. Louis, Missouri
   63110)

Minutes are the official recording of business conducted during a meeting of the society.

News Items are events or announcements of interest to members of the society (meeting dates; promotions, honors, or actions of members).

CHAPTER 14

Solicited and remunerated manuscripts

An editor will often solicit a guest editorial, a review or general survey article, or a book review, any or all of which may be remunerated. Great care must be taken at the solicitation stage if embarrassment and recrimination are to be avoided later. Before issuing the invitation, the editor should be reasonably certain that he will be pleased to publish the manuscript that results, although minor revisions may prove desirable. It is best to make the invitation in writing, so that all conditions and stipulations—deadline, amount of remuneration (if any), maximum length—are clearly recorded. It is also important at this stage for the editor to reserve the right not to publish the article "if it does not meet editorial requirements," or at the very least to reserve the right to make editorial suggestions for revision. If the editor plans to submit such solicited material for review to persons other than himself, he should so advise the author in the initial letter of invitation. Outright rejection of a solicited item is inevitably awkward, but is somewhat less so if the ground is prepared.

Guest editorials and book reviews present the greatest potential difficulty. The editor will naturally want them to meet the journal's standards of literary style, substance, and good taste, and should therefore stipulate his prerogative to offer suggestions for revision. He is not, however, justified in rejecting the manuscripts if they express opinions contrary to his expectations or to his own point of view. The only proper course, then, is to publish the manuscript with a footnote stating that guest editorials (or book reviews) express the authors' views and do not necessarily reflect the journal's policy or position (see "Editorials").

Occasionally, an editor will solicit a research article, to be based on work he knows to be in progress or just to have been completed in a certain laboratory. Such a manuscript is generally expected to be subjected to the same review procedure as unsolicited manuscripts, and to be accepted or rejected accordingly. Again, this stipulation should be made clear in the letter of invitation.

CHAPTER 15

A special type of letter to the editor: comments on published articles

Letters to the Editor are often found in weekly journals with wide scope, such as *Science* and the medical journals. They permit publication of criticism and comment on public issues and on the contents of the journal and provide valuable feedback from readers to editors. Nevertheless, editors generally believe that a section entitled "Letters to the Editor" is not appropriate in a journal devoted exclusively to research articles. We describe here a particular kind of letter which can serve several useful functions in a journal whose main purpose is to publish full reports of completed experimental research.

This sort of letter explains, amplifies, corrects, or otherwise comments substantively upon an article recently published in the journal. Letters of this type are quite distinct from notes and brief communications and can well appear in journals that do not publish notes or brief communications. They are also distinct from unsolicited letters on general topics, or on matters unrelated to the previous contents of the journal.

Such letters to the editor serve several functions. First, persons who feel that their own previous work has been inaccurately cited, wrongly interpreted, or even improperly neglected can say so in a letter to the editor. If the author of the article that is claimed to be at fault serves as one of the reviewers of the letter to the editor, such errors can often be corrected with a minimum of difficulty. Second, such letters allow the correspondent to correct an error or offer an explanation for an observation at less cost to the journal than the publication of an article devoted to the same purpose.

Finally, a reviewer may use this method for making public certain reservations about an article that he has reviewed. In certain circumstances a reviewer may have a fundamental objection that he and the author cannot resolve in the ordinary process of reviewing. The reviewer may nevertheless believe that the article deserves publication. After the article appears, the reviewer may express his opinion, or offer his alternative explanation of the data, just as any other reader may do. He does not, then, identify himself as a reviewer, but does take the responsibility for his criticism by signing the letter. In both these respects, the procedure seems to have advantages over publishing an article with the reviewers' comments appended (an alternative mode of achieving some of the same ends). A further advantage is that the effort required and the formality of the letter provide some assurance that the issue raised will have substance. Many authors find the appending of reviewers' caveats to their articles a distasteful procedure; it can also greatly weaken the pressure on author and reviewer to reach a resolution of the issues during the reviewing process.

Letters to the editor of this kind should be reviewed in exactly the same way as are regular articles. They should be evaluated for accuracy and importance, and need not be accepted unless the reviewers and the editor believe that they warrant publication. In general, one of the reviewers of a letter to the editor should be the author whose article is being commented upon in the letter; he may, of course, respond with a letter of his own, which will be reviewed in turn. There are several advantages to using as one of the reviewers the person whose article is the subject of comment. One is that he is presumably an expert on the subject under discussion. Another is that he may well wish to respond with a letter of his own, and it is a courtesy to allow his response to appear in the same issue as the original letter to the editor. A third is that, if he finds the letter offensive or upsetting, he has an opportunity to make his position clear before publication; this could save a great deal of unpleasantness. An author who is sufficiently offended by the tone or content of a comment on his work may even take legal action; he would be far less likely to do so if he had reviewed the comments before they were published. Moreover, he would probably be unable to take legal action if, acting as a reviewer, he had recommended that the letter be published, especially if the revisions he suggested had been made or if he had seen and approved the letter in the exact form in which it was eventually published.

Such "Letters to the Editor" thus permit any reader to comment upon and criticize articles, and allow reviewers occasionally to recommend acceptance of an article and yet to comment unfavorably upon that article after it has appeared. Even if such letters are received and published infrequently, editors may well find them extremely useful.

To avoid confusion, the editor may wish to entitle these contributions, and the section of the journal in which they appear, as "Comments on Published Articles." The Information for Authors for this section should state that each "Comment" must bear a title and a short abstract for ease of retrieval.

CHAPTER 16

Publication of symposia or conference papers in a journal

The primary reasons for holding a symposium are to provide an opportunity for the thoughtful review of work on the subject and to permit a provocative exchange of ideas. Publication of the symposium, in addition to providing a historical record of what was thought and said about a given subject on a particular occasion, may allow a larger audience to share any benefits from the exchange of ideas. Publication of symposia in journals, however, has certain disadvantages. A participant, for example, may merely iterate previously published information. Another speaker may deliver an excellent presentation that incorporates his latest results, but its publication in the symposium (usually without a detailed description of methods) may preclude more definitive publication later (see "Multiple Publication"). An additional difficulty associated with publication of a complete symposium is the unevenness of the quality of the papers, some of which may be substandard.

The suggestions here may be helpful to the journal editor who has decided that the advantages of publishing a particular symposium or group of papers outweigh the disadvantages or who is obliged by the regulations of the society that sponsors the journal to publish the proceedings of meetings of that society.

LACK OF QUALITY CONTROL IN THE ABSENCE OF FORMAL REVIEW

Since the permanent editor of a journal is responsible for the material that appears in his journal, the key problem in publishing a symposium stems from difficulties in imposing procedures for review and control. The following procedures, although commonly used, do not completely solve the problem.

(1) Stipulation of peer review and the right to reject

In agreeing to publish a symposium that has not yet been organized, the journal editor can stipulate that the contributed papers will be subjected to the customary process of peer review of that journal, and can reserve the right to reject papers that do not meet the journal's standards for publication. This requirement may result in incomplete publication of the symposium, but the editor may, if he wishes, include a list of titles (and perhaps abstracts) of all papers presented at the symposium, along with the names of the participants.

(2) Relegation of responsibility to a guest editor

The editor may shift responsibility for the symposium to a guest editor, such as the organizer of the symposium, but most readers will still consider the permanent editor responsible for the content of his journal. The participation of a guest editor does not therefore relieve the permanent editor of responsibility in the eyes of the readers.

If the guest editor automatically accepts all the papers presented at the symposium, the author of a poor paper that is included for the sake of completeness later acquires a bibliographic reference that erroneously implies its approval by way of the usual peer review of that journal.

(3) Use of distinctive pagination

The editor can agree to publish all the contributed papers, regardless of quality, and use distinctive pagination, such as the addition of S (for symposium) or C (for conference) to the page numbers, to distinguish these papers from the regular, reviewed portion of the journal. Not all readers, however, will interpret the special pagination correctly, even if an explanatory statement is inserted somewhere in that particular issue of the journal.

AVOIDANCE OF DELAYED PUBLICATION OF SYMPOSIA

If the presentations of the participants are to be published, authors should not be required to surrender their manuscripts upon delivery of their presentations, but should be allowed to modify the text, when necessary, according to the discussion and the content of the other presentations. A deadline within two to three weeks after the symposium is reasonable. The organizer of the symposium should clearly announce, two or three times before the symposium is held, that the deadline for submission of final copy will be strictly enforced, with no exceptions. One week before the symposium, a reminder should be sent to all participants. A telephone reminder within a week after the meeting is a sound investment.

EDITING AND PUBLICATION OF DISCUSSIONS OF PAPERS

Although it is neither necessary nor desirable to publish the discussion of each paper in full, some parts of discussions are often pertinent and should be included. The chairman should decide, in consultation with the authors of the papers discussed, which points should be retained and how they might be used. If a discussant requested clarification, documentation, or elaboration of a portion of the paper, for example, appropriate changes can be made in the paper. Another type of substantive contribution by a discussant might be incorporated into a footnote added to the paper, with the author giving credit to the discussant but raising the point in its proper context. Any extended discussion of the paper that cannot be added to the text or inserted in a footnote is best appended in full to the paper, but such circumstances will be rare. The chairman may either arrange for the session to be recorded and transcribed, for later editing by the discussants, or ask the discussants to submit their remarks in written form suitable for publication. The permission of the discussant must, of course, be sought (see "Information for Authors").

Special types of manuscripts

CHAPTER 17

Book reviews

Many journals publish book reviews, some occasionally, others in a regular section in each issue, under the supervision of a special book review staff. Reviews bring to the reader's attention recent publications, and "lead" reviews may alert him to works of special importance.

BOOK REVIEWERS
Selection

Reviewers should be selected on the basis of technical competence, integrity, candor, impartiality, and literary skill. Developing a cadre of competent, reliable reviewers is invaluable to the editor.

Identification to the reader

Since the identity of the reviewer is important in the reader's acceptance of an evaluation of a book, only signed reviews should be published. A note may be included, in the book review section, that, unless otherwise signed, all reviews are by the Book Review Editor, whose name should be identified in the journal.

Remuneration

Most reviewers accept assignments without financial remuneration. Editors generally give the reviewer a copy of the book, provided by the publisher, as token compensation for his efforts.

SELECTION OF BOOKS FOR REVIEW

Journals featuring regular Book Review sections receive numerous unsolicited books from publishers, and careful screening is necessary before the final selection is made by the Book Review Editor, if the journal has one, or by the Chief Editor or an Advisory Editor. The primary criteria for selection should be the relevance and significance of the subject for the readers. Good as well as poor books should be reviewed.

Because of the ever-increasing cost of producing books, publishers are expecting greater accountability from editors for complimentary books sent for review. Some publishers now make a preliminary inquiry to determine if the editor intends to have the book reviewed and some even ask the name of the reviewer and the amount of space to be devoted to the review (*Nature* 248:461, March 29, 1974).

Books received

A list of "Books Received" keeps readers informed of new books and serves as an acknowledgment to publishers of books received by the editor. The editor may wish to indicate how the list is compiled, lest the reader assume that books listed in this section are simply considered unworthy of review. Some books in the list may, in fact, later receive full or brief review.

Reviews

Books considered to be of greatest interest to most readers and of greatest importance to science usually receive full review. Some editors assign books that they consider of lesser import to "Brief Reviews," but such classification obviously involves a value judgment. Because practices vary from one editor to another, it is a mistake to assume that "Brief Reviews" necessarily includes less worthy books. Some books that are actually classics may receive a brief review when a new edition or a reprinting becomes available, without drastic change in content. The length of the review should be the decision of the reviewer, and, indeed, he may write a long critique of a poor book on an important subject in order to point out all the flaws or deficiencies. An editor should therefore avoid adopting the policy of automatically assigning poor books to "Brief Reviews."

INSTRUCTIONS TO REVIEWERS

Most reviewers welcome some guidelines or suggestions for writing a review.

(1) Length

Some journals limit full reviews to 300 to 500 words and brief reviews to two or three short statements. Book review essays, in which the reviewer goes beyond the contents of the book, are, of course, longer.

(2) Scope

A useful review identifies the audience to whom the book would be useful; evaluates the completeness, soundness, and appropriateness of the contents (including the bibliography, illustrations, and index), as well as the readability and literary quality of the publication; and describes the format, typography, and other physical characteristics of the book, when these deserve mention. Reviewers should be advised whether they are expected to write a strict critique of the book itself or to prepare a book review essay, in which they may comment more broadly and philosophically on the subject, including, perhaps, a critique of the general subject. A simple description of the contents of a book is of limited utility, and clichés, such as "The uneven writing in this book reflects the diversity of authorship of the individual chapters," should be avoided.

(3) Date due

As the interval increases between publication of a book and publication of the review, the value of the review to readers decreases. For this reason, prospective reviewers should be asked not to accept the invitation to review a book unless they can submit the review within a reasonable interval—three weeks or so after its receipt. Once a book is mailed to a prospective reviewer, it may be difficult to retrieve it should the reviewer fail to produce the review, and the editor may then have to forego publishing reviews of some important books. He would be wise, therefore, to obtain a commitment for a review, within a specified period, before mailing important or expensive books.

(4) Editorial scrutiny

Reviewers should be advised that solicited reviews will be subjected to editorial scrutiny and that publication is therefore not automatically guaranteed. To minimize biased reviews, the editor may wish, in his letters of invitation to prospective reviewers, to ask those with personal or professional prejudices to disqualify themselves.

(5) Remuneration

Prospective reviewers should be told whether they will receive payment or simply a copy of the book for their efforts.

(6) Identification of the reviewer

Prospective reviewers should be advised that their names will appear under the review.

EVALUATION OF THE BOOK REVIEW BY THE EDITOR

Each review should be evaluated for soundness, objectivity, appropriateness, value to readers, bias, defamatory content, and literary style. The editor should retain the prerogative of withholding publication of unsatisfactory reviews. Before rejecting the review, the editor should, however, identify any weaknesses in the review and try to persuade the reviewer to make the necessary modifications.

LEGAL IMPLICATIONS

A book review should be a candid, critical, fair evaluation, without abrasive, offensive, or libelous statements. Reviews should concentrate on the merits of the book itself rather than on the personality of the author(s). Editors should examine reviews carefully for inappropriate subjective statements, emotionally charged diction, or potentially libelous characterizations, such as "McCarthyism" or "Nazi experimentation." If the editor or his staff carefully scrutinizes the review for any personal attacks on the author, it is unlikely that legal complications will arise (see "Masthead" regarding disclaimer).

EDITORIAL PRACTICES

Chapters 18-31

References

Chapters 21-23

Format

Chapters 25-30

CHAPTER 18

Information for authors

The title "Information for Authors" may be preferable to "Instructions to Authors" for the section that defines the purpose and scope of the journal, as well as provides specific information about content and format of acceptable manuscripts. Ideally, this material should be placed in the same, easily found place in each issue and, for ready accessibility, should be listed in the table of contents. Although it cannot be a complete guide to good writing practice, it should define specific editorial policies and practices that the editor wants prospective authors to observe in the preparation of manuscripts for publication. The information should be succinct; lengthy instructions only discourage reading and bewilder authors.

The following items include various kinds of information that editors may want to convey to prospective authors. An editor's selection of items will be dictated by the nature and requirements of his particular journal; he will obviously omit items that have no relevance to his journal. Since a prospective author can gain a great deal of information by consulting a recent issue of a journal, some editors may want to provide only the information that cannot be obtained in that way.

GENERAL MATTERS
Purpose and scope

The title of the journal may make the purpose and scope obvious, but the editor may wish to specify the topics that are automatically excluded and those that appear only rarely in the journal, particularly if it is a new journal. Such information is useful to prospective authors.

Types of manuscripts considered

The editor may wish to specify, and perhaps define, the types of communication that will be considered for publication, such as research reports, analyses of clinical cases, preliminary communications, brief communications, review articles, letters to the editor, or case reports. If only solicited editorials are considered, a statement may be made to this effect.

Length of articles

Text. Any limitation imposed on the length of the text, on the number and type of tables, graphs, photographs, or other illustrations, or on the number of references should be stated.

Page charges. Page charges levied against authors should be specified, with an indication of whether the payment of such charges is a prerequisite for publication.

Numbered series

If numbered series of articles are discouraged, the editor may wish to inform authors that readers may be guided to earlier related papers by means of footnotes, by a statement in the introduction, or by suitable use of references.

Reviewing policy

The editor may inform potential contributors that the acceptability of manuscripts is determined by review by one or more reviewers if that is the policy of the journal. He may wish to indicate the usual time required for advising an author of the editorial decision and the usual interval between acceptance and publication (see "Publication of Dates for Manuscripts as an Index of Priority").

Accelerated publication. The editor may wish to state the conditions under which certain types of papers receive accelerated reviewing and publication, and may give instructions for submitting such papers.

Prior or multiple publication (see "Multiple Publication"). Although submission of a manuscript usually implies that it has not been published or submitted for publication elsewhere, in whole or in part, the editor may wish to require that the author make an explicit statement to this effect. If so, the editor should define "prior or multiple publication," as, for example, "any form of preliminary publication other than an abstract of not more than 400 words," and should specify that a copy of any earlier, or contemplated, publication should be provided for the editor's consideration.

Ethical experimentation (see "Ethical Experimentation and the Editor"). The editor may wish to request authors to provide (1) a photocopy of approval of the research by the institutional committee on ethics or (2) assurance that experimental animals have received proper treatment and care, or should otherwise advise authors of the editorial policies regarding ethical experimentation.

Revision. Although some editors specify, in Information for Authors, that authors who submit a revised manuscript are expected to reply to the reviewers' comments point by point, this information is more effectively provided in the letter to the author in which revision is requested.

Rejection. The editor may wish to state that his decision is final.

Priority claims and promissory notes

The editor may wish to state that priority claims and promissory notes, such as "This is the first report of . . ." or "Experiments on . . . are underway (or contemplated)," are unacceptable. Prohibition of priority claims in titles may be similarly specified.

Copyright and other legal considerations
(see "Copyright")

The editor may wish to state explicitly that a manuscript, once accepted for publication, becomes the sole property of the journal and that permission must be obtained to reproduce it, in part or in whole. In such cases, it may be well to identify the actual owner of the copyright (the journal, the publisher, or the society that sponsors the journal). Some journals state that abstracts may be reproduced without specific permission, provided acknowledgment of the source is made.

Editors of clinical journals, especially, should advise authors that patients must be unrecognizable in photographs unless specific written consent has been obtained, in which case a copy of the authorization should accompany the manuscript. Merely blocking out the eyes of patients is not sufficient to mask identity.

The editor may wish to advise authors that they are responsible for all statements in their papers.

Some journals contain a statement that the editor does not assume responsibility for the loss of manuscripts.

Address for submission of manuscripts

Prospective authors should be given the name and address of the person to whom manuscripts should be submitted. It is wise to provide this information *after* important statements on policy.

Reprints

A statement may be included about whether the journal provides authors with complimentary reprints, or where and how they may be ordered.

FORM AND STYLE

Many rules governing form and style are adopted for the convenience of the editorial and redactional offices and the printer. Inclusion of minutiae about such matters risks the author's disregarding more important instructions for preparing his manuscript. For this reason, editors may find it desirable to have a separate section for mechanical aspects of the preparation of the typescript, or may advise the author to consult recent issues of the journal for guidance.

Title

The importance of succinct, informative titles should be emphasized, and any limitation on the length of the title should be specified.

Running title

If a running title is required, indicate any limitation on the number of characters, including spaces.

Key terms

Authors may be asked to list key words or phrases that help alert the reader to topics in the paper that are not referred to, or implicit in, the title. Such key words are often used by abstracting and indexing services (see "Abstracts" and "Indexes").

By-line (see "By-lines and Acknowledgments")

Editors may wish to advise that only the names of those who have contributed materially to the preparation of the paper should be included in the by-line. Authors should be asked to identify the person to whom reprint requests or other correspondence is to be directed.

Acknowledgments (see "By-lines and Acknowledgments")

Authors should be advised to obtain, from the person(s) thanked, approval for the content and wording of acknowledgment of assistance.

Equipment and supplies

The editor may wish to instruct authors to provide the names and addresses of manufacturers and suppliers of special material and equipment.

Abstract (see "Abstracts")

Abstracts should be informative rather than descriptive, intelligible when divorced from the article, devoid of undefined abbreviations, and suitable, without rewriting, for reproduction by abstracting services. Any limitation on length should be specified. If the journal also requires a summary at the end of the article, the difference between the abstract and the summary should be specified.

Organization of article

State any requirements or preferences regarding organization. A rigidly prescribed form is not recommended.

Illustrative aids

By specifying any restrictions on type and number of illustrations, and by indicating the size of illustrations and of lettering required for legibility after reduction, the editor may avoid redoing such material after the article has been accepted. Authors should be requested not to use paper clips on illustrations, since the indentation they make may show on reproduction. A preference for glossy prints should be stated. Letters granting permission to reproduce or modify previously published illustrative material may be required. The editor may wish to discourage duplication of material in text and in illustrative aids (tables, graphs). A title should be required for each table, and a legend and title should be required for each graph, photograph, spectrum, drawing, or other illustration. Some editors request that authors indicate the preferred position of illustrations in the text.

(1) *Tables.* Authors may be instructed to make tables simple and intelligible without reference to the text and to provide them with succinct, informative titles. The method used for keying footnotes to tables should be specified (symbols or superscript letters).

DEPOSITION OF DATA. The editor may offer to deposit oversized tables of data in a national depository in lieu of publication, if the article is otherwise acceptable.

(2) *Graphs.* The physical form in which graphs are acceptable should be stated, for example: original drawings or glossy prints; black on white or colored inks. Specifying the maximum number of bars that the editor permits to be superimposed on a graph or the maximum number of curves included in a graph may discourage cluttering.

(3) *Halftones, including roentgenograms, photographs, photomicrographs, electron micrographs, scans.* Instructions may be given to crop photographs to display only the field of interest; to include a measured bar, when necessary, to show scale; and to indicate magnification in the legend. When several illustrations are grouped in a composite figure, identification of parts (A, B, C, . . .) should be included in the illustration. Whether mounting of figures is desired or unacceptable should be indicated.

(4) *Electrical tracings, such as electrocardiograms, electroencephalograms, ultraviolet spectra.* The editor may wish to note whether original tracings are required rather than line drawings by an artist.

(5) *Color plates.* The conditions, if any, under which color plates are acceptable should be stated. The editor may wish to recommend that authors submit original transparencies for colored illustrations, since black-and-white reproductions of colored illustrations are often unsatisfactory.

Statistics

Authors should be required to specify whether they are citing standard deviations or standard errors of the mean and to indicate the number of observations involved, the tests of significance used, and the appropriateness of these tests.

Nomenclature

Reference should be made to the authority followed by the journal for nomenclature, abbreviations, and units in appropriate fields. Any policy on the use of trivial *versus* systematic names of chemical compounds and generic *versus* trade names of drugs should be defined. The Système International (SI) for all units should be recommended, with any exceptions that are acceptable to the journal being specified, and authors should be encouraged to record at least some of their values in the form actually observed, as well as in derived or recalculated form.

The editor may wish to give instructions about such mechanical details as whether numbers are to be spelled out or recorded in Arabic numerals.

Mathematical symbols and equations

Authors should be asked to identify all handwritten symbols and to list all special symbols used. They should also be encouraged to simplify notations for ease of typesetting (for example, to use fractional indices instead of root signs, and "exp" instead of "e" raised to complicated powers). For setting of mathematical material, the editor may wish to refer authors to: Chaundy, T. W., Barrett, P. R., and Batey, Charles: *The Printing of Mathematics*. London, Oxford University Press, 1954.

References (see individual chapters on references)

The method used by the journal for citing references within the text and at the end of the article should be described. "Unpublished work" and "Personal communication" should not be permitted in References Cited but should be incorporated in the text. Authors should be instructed to provide letters of permission from persons thus cited. The author's responsibility for verifying references against the original sources to ensure accuracy should be emphasized.

Textual footnotes

If footnotes are allowed, a statement should be made to advise authors to keep them to a minimum and to distinguish them from References Cited.

PREPARATION OF TYPESCRIPT

Number of copies

The number of copies of the manuscript and of illustrations required by the journal should be specified. The editor should specify whether carbon, electrostatic, or lithographic copies are acceptable.

Spacing

Double-spacing should be required throughout the manuscript, including tables, footnotes, quotations, legends, and references. Width of margins desired should be specified.

Pagination

Any requirements should be specified for separate pages for title, by-line, authors' affiliations, acknowledgment of financial support, and abstract. Preferred order of the pages of a manuscript should also be specified. Many journals require that each page of the manuscript be identified by at least the name of the senior author and a running title.

Tables

Inclusion or exclusion of vertical rules and other special features required for tables should be indicated. Double-spacing should be required for the footnotes, which the editor may require to be typewritten on a separate sheet.

Legends

Legends for each illustration should be typewritten, double-spaced, sequentially on a separate sheet. The title of the illustration should be given in the legend, not in the figure itself.

Illustrations

Directions should be given for identifying each illustration by means of a label pasted on the bottom of the reverse, and for indicating "Top" on the reverse, when necessary.

References

The specific style used by the journal for citing references should be described in detail for journal articles, books, chapters in books, and technical reports. The importance of double-spacing references on a separate sheet should be emphasized.

GALLEY PROOF

The editor should indicate that only minimal changes should be made on galley proof, that the cost of excessive changes will be charged to the author(s), and that unless galley proof is returned within a specified time, the editor reserves the right to publish the manuscript as it appears on galley proof.

CHECK LIST*

Some editors include a check list for authors to use to make sure they have included all necessary material, in proper order, when they submit their manuscripts. A sample check list might read as follows:

Letter of submission
Original typescript of article (double-spaced throughout)
  Title page
    Title of article
    Full name(s) of author(s)
    Affiliations of author(s)
    Author to whom correspondence is to be sent
  Abstract (double-spaced)
  Article proper (double-spaced)
  References (double-spaced)
  Tables (double-spaced)
  Legends (double-spaced)
  Illustrations, properly labeled
Permission to reproduce published material or cite unpublished data
Key words for index
Additional copies of manuscript, as required by journal

CHAPTER 19

Copyright

Copyright protection covers the particular way in which an author has expressed himself, in word and format, about a given problem, but does not extend to ideas, systems, or factual information that his work may convey. The Federal copyright law (Title 17, U. S. Code) grants a copyright owner certain rights, infringement of which entitles him to monetary damages and injunction. These rights include, among others, the exclusive right to print, reprint, publish, copy, translate, and vend the copyrighted work. A copyright is obtained by sending two copies of the work to the U. S. Copyright Office in Washington, D. C., along with an application and the fee. The copyright notice must be affixed to each copy of the work published.

Many persons do not observe the strict limitations imposed by copyright law, but take advantage of the rubric of "fair use" of copyrighted work. In the broadest terms, the doctrine of "fair use" means that in circumstances in which the use is reasonable and "not harmful to the copyright owner's rights," copyrighted material may be used to a limited extent without obtaining permission. Under this doctrine, for example, scholars, critics, and libraries have been free to publish, without permission of the copyright owner, short extracts or quotations from copyrighted works for the purpose of illustration or comment.

The distinction between "fair use" and unlawful infringement is not easily defined. There is no specific number of words, lines, or notes, for example, that can safely be duplicated without permission. Merely acknowledging the source of the copyrighted material does not avoid infringement.

It is still mandatory to get permission from copyright holders before incorporating copyrighted material—particularly charts, graphs, photographs, illustrations, excessively long quotations, or full articles—into papers or textbooks to be published. It is ethically desirable, and customary, for the editor, publisher, or copyright holder to grant such permission contingent upon the author's approval and to send a copy of such conditional permission to the author. The specific use for which the material is intended must be expressed in the letter of request. Some copyright owners require a fee for the use of copyrighted material in their possession.

The legal implications of modern technologic advances in high-speed reproduction of portions or complete articles and booklets were tested in the case of *The Williams & Wilkins Company against The United States.* After more than seven years of litigation, this case was resolved by a split decision by the U. S. Court of Claims, which ruled that photocopying of magazines and books by scientists and the making of single copies for individual users by libraries does not constitute infringement. This ruling represents a broad interpretation of the "fair use" doctrine, but does not abrogate any existing copyright statutes. The case was appealed to the U. S. Supreme Court, which, owing to a tie vote in February, 1975, failed to rule on the case. The tie vote meant that the decision of the Court of Claims holds. Publishers, librarians, and photocopy manufacturers are now trying to work out an equitable system for the fair use of copyrighted material.

Under the concept of "fair use," photocopying of single articles for personal, noncommercial purposes (education, research) seems entirely proper, at present. Realistically, there is no practical way to control photoduplication of single articles or chapters by individual users, so widespread and accessible is the equipment.

Mass photocopying, however, such as duplication of entire journals or books, or of large portions thereof, is generally regarded as an abuse. Reproduction of material for educational purposes (by librarians and educators for limited classroom use, by investigators or by students within reasonable limits) is often differentiated from that for profit. If, however, "for educational purposes" is interpreted to mean multiple copying of book chapters or papers for successive series of large classes or for incorporation into an informal "book," this may be construed as abuse of "fair use." When a group of published articles is photocopied, collected, bound, distributed, and sold as a separate volume, even if at no profit to the collector, the "fair use" mandate has been abused. Such practices have economic implications for the author and pose a significant financial threat to journals with small circulations.

The "fair use" concept remains ill-defined. There is active Congressional interest in redefining copyright law, in keeping with the expansion of publishing practices. Most courts, however, still make an effort to distinguish between "fair use" and "infringement" on the basis of the Register of Copyrights Report of 1961. These criteria are:

(1) purpose of use,
(2) nature of the copyrighted work,
(3) amount and substantiality of the material used in relation to the copyrighted material as a whole, and
(4) effect of the use on the copyright owner's potential market for his work.

It is item (4) that carries the greatest weight with judges and juries.

The copyright policy of the National Institutes of Health, as defined in the *Federal Register* **38**(197):28314-28315, October 12, 1973, indicates that: "Except as may otherwise be provided under the terms and conditions of a specific grant award, the grantee may copyright or arrange for copyright, without prior National Institutes of Health approval, any publication, film or other similar communication material developed or resulting from a project supported in whole or in part by a National Institutes of Health grant. Such copyright, however, is subject to a royalty-free, nonexclusive and irrevocable license or right in the government to reproduce, translate, publish, use, disseminate, and dispose of such material and to authorize others to do so. In addition, communications in primary scientific journals publishing initial reports of original research supported in whole or in part by the National Institutes of Health may be copyrighted by the journal with the understanding, however, that individuals are authorized to make, or have made by any means available to them, without regard to the copyright of the journal, and without royalty, a single copy of any such article for their own use." Such material, according to the *Federal Register,* is in the public domain and is not "owned" by the journal concerned. The implications of this definition, insofar as existing copyright law and the new pronouncements on "freedom of information" are concerned, have not been explored. Some may regard it as a copyright violation and even intrusion by the government into freedom of the press.

Inclusion of a statement in each issue of a journal that reproduction requires permission of the copyright holder should have a salutary effect (see "Masthead"). Editors are advised to seek the advice of legal counsel familiar with copyright law or to consult with their publishers when confronted with specific copyright problems.

CHAPTER 20

Publication of corrections (errata)

From time to time, the editor of a journal or an author wishes to correct a typographic error or a published statement or arithmetic value that was later perceived to be erroneous. Most editors have been content to publish a notice, usually headed Erratum, and leave it at that. Readers and librarians are supposed to make the necessary annotation on the original article, but they often fail to do so. Only in a few journals is the erratum sufficiently well indexed to ensure that someone searching the archives at a later time will be guided to the erratum when he finds the original article, and hence will be enabled to make the correction for himself if necessary.

We recommend the following procedures for bibliographically coupling a correction to the article containing the error, to facilitate the reader's making the corrections when they are first published. In the following discussion we have used *Corrections* rather than *Errata,* both because English terms are generally preferred to Latin ones in journals published in English and because the term *Corrections* can include the addition or substitution of information by the author in the light of new knowledge and need not be confined to announcements that a mistake has been made. The following suggestions and recommendations, however, apply to the handling of errors under any name preferred by the editor.

SUGGESTIONS

(1) The corrections should be displayed prominently, not placed inconspicuously among less important material.

The correction notice can be enclosed in a box or surrounded by a large white space, preferably with a bold heading in which the title of the article or the authors' names catch the eye. If for any reason the correction must appear in a small space, conspicuous typography can be used to call attention to it. The temptation to make as little as possible of a published mistake should be resisted.

(2) Corrections should, if possible, appear in the same place in every issue, close to some frequently consulted portion of the issue.

Suggested places are on or opposite the contents page, on the title page, opposite the inside front cover, opposite the Information for Authors, opposite or below the issue index, if there is one. Above all, the notice should *not* be placed on a page that may be discarded when the journal is bound (see "Binding Practices").

(3) Attention should be drawn to corrections by prominent entries in the list of contents of the issue. Each entry should state the title of the article to be corrected.

Whether the correction is published near the contents page or not, it should be listed in the contents. The signal should not simply take the form "Correction, p. 276," because this is uninformative to the reader of the journal and useless to the scanner of *Current Contents*. The entry should, rather, read:

Correction to Jones et al. Circadian Rhythm in Moon Spores (**46**: 129-137, 1968) . . . . . 276.

(4) In the correction notice, full bibliographic details of the original article should be given. The text containing the error should be printed next, and then the corrected version. Adhesive strips may be used *in addition* to the notice, to make the task of correction easier.

If printing both faulty and corrected texts is too costly or cumbersome (as in the case of a table), the editor may, instead of using the recommended form, give exact instructions for making the correction.

The adhesive strip and the correction notice serve different purposes; the strip should not be substituted for the notice. If the slip becomes detached and lost, the situation is irremediable unless the notice retains the full details.

An adhesive strip that is simply and quickly applied is one that is designed for the title page of the article containing the error, and that reads "See Correction, vol. 46, p. 276." When the article is consulted in the future, even as a photocopy, this marking will bring the correction to the reader's or librarian's notice even if the text itself has not been changed.

(5) Corrections should be thoroughly indexed in the volume indexes. Sometimes a separate Correction Index for the volume is justified: if so, attention should be drawn to it under the headings "Corrections" and "Errata" in the subject index; if not, the subject index should have an entry "Corrections" (cross-indexed also under "Errata") under which every correction is listed. Ideally, correction entries should also be made in the subject index under all subject headings used for the original article and in the author index under every author entry for the article to be corrected.

(a) In the author index, an entry should be made for each author of the article to be corrected: "Smith, V. P. (Correction) . . . . . 276." If the article and the correction have been published in the same volume, the entries can be combined thus: "Smith, V. P. . . . . . 129-137 (Correction, 276)."

(b) The subject index is more complicated. If the article and the correction have been published in the same volume, each entry should carry both page numbers, for example, "Circadian rhythm: In moon spores . . . . . 129-137 (Correction, 276)." If the correction refers to an article in a previous volume, the main subject index entries for that article (words in the title) should be inserted afresh in the form: "Circadian rhythm: In moon spores (Correction to Jones et al. **46:** 129-137) . . . . . 276." This may seem cumbersome and expensive, but the rewards to the searcher are great.

(c) If a complete Correction Index is provided for a volume (either as a separate index or as an entry "Corrections" in the subject index), the most useful arrangement for readers is probably alphabetical by name of first author.

(d) If the journal customarily prints a volume contents, in which the lists of Contents for each issue are reprinted in chronologic order, the Corrections should not be deleted when the volume contents is made up.

(6) When conscientious authors cite an article that they know has subsequently been corrected, they add to their bibliographic citation of the article a citation of the correction.

(7) When corrections are embodied in a letter to the editor, the editor should take care to signal them both on the contents page and in the index in order that they may be coupled to the original article.

If the letter to the editor corrects a mistake in an article by a different author from the author of the letter, the editor will, of course, consult with the original author about the propriety both of publishing the letter and of signaling it as a correction.

References

CHAPTER 21

References cited: recommended form

Not all journal editors require the title and inclusive pagination of a journal article to be included in the list of references. Although these elements are not strictly necessary to enable the reader to find the cited article, we believe that in research reports as well as in review articles, titles and inclusive pagination for references cited provide valuable information to the reader and that the author should therefore provide them.

ADVANTAGES OF USE OF TITLES

(1) In a research report, articles are often cited in support of a single, specific point, and some may therefore argue that the title of the cited article is unnecessary. For the reader well versed in the subject, this may be true, but for other readers, the title of the article may indicate whether the point being documented by the reference is or is not the main subject of the cited article.

(2) A reader unfamiliar with the publications related to the article he is reading may, by inspection of titles cited (particularly if they are sufficiently accurate), be able to select articles that will extend his knowledge of particular aspects of the subject. Such use of References Cited underscores the importance of the editor's attention to the accuracy of titles of articles.

(3) A reader who is fairly familiar with related publications will often be reminded by the title that he has read the article cited, whereas the authors' names and bibliographic data alone may not recall the article to his memory. Further, an author or a group of authors often publishes several papers in the same journal in the same year; the title then clarifies which paper is being cited.

(4) The title will often enable the reader to find the article cited even when the reference contains numerical or spelling errors.

(5) The necessity to provide the title of a reference may require the author to return to the original source, and he may thus discover that his

bibliographic data were incorrect. Merely reading the title may remind him that the paper dealt with aspects of the topic he had forgotten about, and may suggest the wisdom of reading the paper again. He may even realize that he has selected an inappropriate article for citation.

(6) The title of an article in a foreign language automatically indicates in what language the article is written. (Titles in languages that do not use the Roman alphabet should be transliterated or translated, with a notation of the original language in parentheses.)

ARGUMENTS AGAINST USE OF TITLES

(1) Some editors may be concerned that authors will object to the effort required to ascertain or verify the titles of articles. We believe, however, that authors should understand that this is *their* responsibility and that carelessness in this portion of the article may reflect general carelessness in other aspects. Moreover, editors of journals that have recently adopted a rule requiring titles have reported that there were remarkably few such complaints.

(2) Managers of journals may be concerned about the extra space required by references that contain titles. The necessity for extra space can be avoided by the use of a smaller typeface for the references: 7-point, or even 6-point, type is adequate.

(3) Extra expense in composition cannot be gainsaid and must be weighed against the advantages listed here.

ADVANTAGES OF INCLUSIVE PAGINATION

(1) Provision of inclusive pagination often enables the reader to distinguish a definitive article from a preliminary communication when he knows that the work has appeared in both forms, and from a brief announcement or abstract, particularly if the abstract is so identified in the reference.

Apart from the scholarly value of this information, it may have important practical application in helping the reader to estimate the cost of duplication or translation of the cited paper. Librarians request this information when articles must be obtained through interlibrary loan.

(2) Provision of the closing page number may facilitate retrieval of the article cited if the initial page number is in error.

ARGUMENTS AGAINST INCLUSIVE PAGINATION

The effort required to ascertain the inclusive pagination of an article is minimized if the author adopts the habit of recording a reference in full when he initially consults a journal article. Arguments on the basis of extra space and composition necessary are inapplicable, since both are negligible.

CHAPTER 22

References to unpublished or inaccessible information

It is poor practice to allow authors to mix, in the list of references, citations of published documents with citations of "Personal communication,"
"Unpublished data" of the author, and work "To be published." The reader
who encounters a citation number (or name) in the text is misled—even if
only temporarily—into thinking that the statement made is supported by
published work, and may therefore give it more weight than it deserves.
Citations of unpublished material or information of limited availability fall
into the following categories:

(1) "In press"

An article that has been *accepted* for publication, but has not yet been
published, constitutes an acceptable reference. The name of the journal must
be specified, and it is reasonable to require the author to submit a copy of
the manuscript along with the manuscript under consideration.

(2) "Submitted to Journal X"

Some editors permit this as a reference, on the grounds that it provides
the reader with more information than "To be published." We recommend
that "To be published" be substituted, and the recommendation in (3)
applied.

(3) "To be published"; "Manuscript in preparation"

These designations give some useful information to the reader, but the
first is to be preferred to the second (especially as the second often leads to
a listing of all authors of the promised publication, which often lends a quite
spurious air of precision to something in the future). We recommend that
allusions of this kind be given in the text as "(data) to be published," not
included among References Cited.

(4) "Unpublished data"

This designation signifies that the author has no immediate intention of publishing the data, although they may be available to interested persons on request. This kind of citation should be handled in the manner described under (3).

(5) Reference to documents with limited circulation

If documents with limited circulation (lengthy descriptions of procedures that have not been printed, but are available on request or technical reports that are available only by a special procedure) are cited as a reference, the author should provide full details of how copies may be obtained.

(6) Reference to documents that are not available on request

If the author wishes to give credit for the source of ideas or methods by making reference to a document that is not available on request (for example, one with circulation restricted by government or company regulations), the reference should contain a statement to this effect, and the text should be written in such a way that the reader is given all the information he needs without having to consult the classified document. If this would violate the intent of the restriction, neither the reference nor the information in it should be used in the unrestricted publication.

(7) "Personal communication"

Authors may want to refer to the unpublished work of others that they have heard about at a conference, in conversation, or in correspondence. Any such citation should be handled as in (3). In addition, we strongly advise that the editor insert a statement such as the following in his Information for Authors: The author submitting an article should state in a covering letter that the content and wording of references to unpublished work of others have been approved by the person(s) whose work is quoted. This assures the editor that the person cited is not being taken advantage of, that the author has not misunderstood the information, and that the person cited has not changed his opinion about the work referred to on the basis of further work or thought.

The editor can further serve the reader by requiring that the author provide the current address of the person cited. Some editors require the author to furnish a written statement from the person cited, stating that he approves the actual wording of the citation of his unpublished findings or personal communication. This may be achieved by asking that the person cited write his permission on a photocopy of the pages of the text that contain views attributed to him and that those pages be sent to the editor. Editors who follow this policy find that the person cited often requests notable changes in the passages in which his views are presented.

References

CHAPTER 23

Editorial responsibility for verification of bibliographic references

A published list of references frequently contains errors, some serious enough to prevent a reader's finding the paper referred to, others serious enough to lead to time-consuming searches before the paper is found. Although the policy of some journals is that the accuracy of references is solely the authors' concern, it can be argued that the editor who is serious about serving his readers should assume the responsibility of having references verified at the journal's expense.

The thought of the expense will be most editors' primary deterrent, but this need not be excessive. In one journal for which we have data, the cost was $600 per year (240 hours at $2.50 per hour) for an 800-page volume with about 150 articles and 2,400 references. This amount represents 0.9% of the total operating budget. The checking resulted in the correction of major errors (which would have impeded retrieval of the reference, though not necessarily prevented it) in 3.5% of the references and the correction of minor errors (incorrect initials or spelling of a junior author's name) in another 13%. We believe that the result justifies the expenditure. Actually, the expenditure may be negligible. At least one journal has for years made the verification of references a part of the regular duties of two editorial assistants.

Another deterrent is the editor's intuitive feeling that a reference-checking system will not ensure that *every* reference is correct. Even if his reference-checking assistant has access to a major library, a certain proportion of the sources will not be available, either because the library did not purchase them or because they are lost, borrowed, or at the bindery. In the sample from the journal referred to, 10% of the references were unavailable to the checker. If we assume the "major error rate" in these references to be 7% (twice as high as in the "found" sample), this would give us a further 0.7% of the total references containing major errors. Given that these remain uncorrected, it still seems to us that 99.3% correct, while far from ideal, is a substantial improvement over 95.5%. (Naturally, this is

89

only one sample and may not be typical; but it tallies well with a general impression among journal readers that the proportion of errors is surprisingly and irritatingly high.)

Who is sufficiently qualified to check references at a reasonable cost? Particularly fastidious laboratory technicians, young copy editors or others trained in keeping detailed records in publishing operations, librarians' assistants, and skilled typists may be willing to spend a little extra time in their customary working environment to earn some pocket money, and each proved satisfactory during ten years' experience with the journal referred to here.

CHAPTER 24

Copy-editing

Once the editor has accepted an article for publication, it is sent to the copy-editing office, or "redactory." Copy-editing is the process of correcting and preparing a manuscript for typesetting and printing. Some journal editors have copy editors within their own offices, under their direct jurisdiction, whereas others use the publisher's, or some other commercial, redactor. In some instances, free-lance copy editors perform this function for journals. Since the reputation of a journal is affected by the quality of the copy-editing, anyone considering an editorship will want to investigate the administrative relation between the editor's office and that of the redactor to determine the degree of control the editor has over this aspect of journal publication. If copy-editing is within the editor's jurisdiction, he will want to establish adequate educational requisites to ensure competence of his copy editors.

Reviewers who evaluate manuscripts for publication focus primarily on scientific content and only occasionally make grammatical or stylistic changes on the typescript, although they may note, in their evaluations, that the literary form requires some improvement. If the reviewer considers the writing so poor that the meaning and validity of the manuscript cannot be determined, he will recommend returning the manuscript to the author. If, on the other hand, the reviewer deems the content worthwhile but the narration in need of improvement, the editor may (1) return the manuscript to the author for such revision before acceptance or (2) instruct the redactional staff to make the necessary improvements. The choice is usually based on the extent of revision needed to meet the journal's standards.

The amount of copy-editing done on manuscripts varies from one journal to another, ranging from mechanical changes for consistency to removal of grammatical infractions or even to extensive revision that may verge on rewriting. Some journals, and some redactional manuals, contain a statement that the copy editor handles only those manuscripts that have been judged scientifically acceptable and accurate, and that he is therefore not expected to do professional rewriting or to rectify errors in fact. Only the rare scientific manuscript requires no changes at all. Minimal changes consist in making the manuscript conform to the editorial practices of the journal regarding spelling, symbols, abbreviations, hyphenation, capitalization, punctuation, nomenclature, format for citation of references, and other mechanical considerations, which may be detailed in the copy-editing manual of the journal or publisher. Many journals prefer English to foreign plurals (*atriums* instead of *atria, hemangiomas* instead of *hemangiomata*) and have other usage preferences, such as *kill* instead of *sacrifice* and *roentgenogram* instead of *x-ray*. Some copy-editing will undoubtedly be done mechanically in the future. With the development of electronic devices capable of character recognition, computers may be programmed to convert British to American spellings and foreign to English plurals, and to make other such changes that do not require intellectual judgment.

The copy editor makes certain that approval has been obtained for repro-duction of copyrighted material, that the type of stain and the magnification of photomicrographs are specified in legends, and that a credit line is in-cluded in the legend, when necessary. He also matches the textual references to tables, figures, and bibliographic citations with the actual parts appended to the manuscript, to make sure they are all present and in the proper order. Some copy editors are expected to verify references, at least on a spot-check basis. In addition, the copy editor may write marginal queries to the author regarding ambiguous passages, seeming contradictions, illogicalities, or de-ficiencies, such as incomplete references or omission of the abstract or legends. Finally, he may mark the manuscript for the printer, indicating spacing, the position for insertion of tables and figures, and the size and face of the type to be used for the various parts—title, headings, running heads, underscored words, block quotations, case histories, references cited, tables, legends.

After the manuscript has been copy-edited, it may be sent either to a senior scientific editor for approval or to the production office. Some editors send the edited typescript to the author for approval before it is set in type. When the galley proofs have been printed, they are sent with the manuscript to the copy editor (or proofreader), who will correct obvious errors, answer the compositor's queries whenever possible, and make certain that the queries to the author have been transferred to the galley proofs. The galley proofs are then sent to the author, who will correct errors, including any inap-propriate changes made by the copy editor or others. In some journal offices, galley proofs are sent simultaneously to the author and the proofreader, in the interest of saving time. The author is often supplied with instructions for proofreading, including proofreaders' marks, and may be asked to make only essential changes, since resetting in proof is costly. Some publishers, in fact, charge authors for any proof changes from the original text beyond a reasonable and specified number. The author is given a deadline for returning the galley proofs, usually about five days, after which the material will be published as it appeared on the galley proofs.

When the author returns the galley proofs, the copy editor (or proof-reader) reviews the author's corrections, making sure they will be clear to the compositor, and sends them to the production office. Should controversies occur between author and copy editor, a senior editor usually arbitrates. Once the printer makes the corrections noted on the galley proofs, a copy editor (or proofreader) should check them carefully on page proof, to make certain that new errors have not been introduced in the resetting. Page proofs are usually examined by the journal's copy editor only and are not seen by the author of journal articles.

COPY-EDITING PROBLEMS

Elective or capricious changes in wording that result from personal usage preferences of the copy editor are to be discouraged. Changes should be made with discretion and common sense. Obsession with minutiae and pre-occupation with prescriptive rules of grammar may block the judgment of some copy editors regarding the appropriateness of their changes in the author's context. Prejudices of the inadequately trained can create serious, and even ludicrous, distortions of the author's meaning. An author may be so annoyed by changes that mutilate his text or introduce errors into it that he will refuse to send any future manuscripts to the offending journal, and may advise his colleagues to avoid publishing in that journal.

Copy-editing is best limited to essential changes that can be fully and indisputably justified. For one thing, copy editors are often not scientifically trained and may inadvertently alter meaning when they make changes in wording or punctuation. For another, heavily edited material eliminates the natural variation in literary styles among authors, and the resulting stereo-type makes for tedious reading. For a third, heavy copy-editing is costly. When, therefore, a manuscript is so poorly written as to require major revision or rewriting, the editor is wise to return the paper to the author for recasting. Such a practice, if adopted by all journal editors, might have a salutary effect on the quality of scientific publications.

Since the person whose name appears in the by-line is the one held publicly accountable for the article, he should not have the literary tastes or idiosyncracies of another imposed upon him. An editor, on the other hand, has the prerogative of refusing to publish an article that he thinks violates the literary standards of his journal. A reasonable editor and a reasonable author, however, can usually arrive at a satisfactory resolution of differences regarding editorial changes in the author's manuscript.

Format

CHAPTER 25

Journal cover

Basic items on a journal's cover that help readers, librarians, and secondary services are given in the American National Standard for Periodicals: Format and Arrangement.[1] We comment on some of these items and add others.

(1) Title of the periodical, volume number, issue number, date of issue (preferably month and year). Optional: name of the society sponsoring the journal.

(2) Inclusive pagination (usually on the spine, together with date, volume number, and issue number. The spine should be printed in such a way as to be readable from left to right when the issue is lying flat, front cover uppermost[1]; in other words, from top to bottom when the issue is shelved.)

(3) On an issue containing an index, the words "Index Issue" (preferably on the spine). Practices vary widely among different journals as to which issue contains the volume index(es); the recommended addition to the journal's cover helps readers, librarians, and bibliographers, who otherwise spend much time searching for the index issue of an unbound volume.

(4) Table of contents. Readers dislike having to search for the table of contents, especially if it is buried among pages of advertisements. Contents on a cover (whether front or back) are immediately obvious.

One argument advanced against this recommendation is that the cover may be lost or mutilated in mailing. The Standard[1] suggests printing the contents inside as well as on the cover to avoid this problem, but this is too expensive for many journals. If such mutilation occurs often enough to be a serious consideration, the method of mailing the journal should be reexamined.

Another argument against the recommendation is that the contents of the issue printed on a cover are lost when the issues are bound. This is usually no great loss. If a volume contains a large number of titles, a cumu-

94

lated Volume Contents is too cumbersome to be useful and readers prefer a subject index. If, on the other hand, each volume is small and a Volume Contents therefore useful, this can be cheaply provided in inside pages by printing from type kept standing from previous issues or by photo-offset. Furthermore, if for special purposes a subscriber or library wishes to preserve lists of contents that appear only on covers, these may in fact be bound together with the rest of the journal.

We believe that the advantages of printing contents on the cover far outweigh the disadvantages. If the list of contents is too long to fit on one cover, it may be continued on as many inside pages as necessary.

(5) Corrections (Errata). We recommend that Corrections be prominently announced on the contents page.

(6) Reference to the Information for Authors. If the Information for Authors appears anywhere but on an inside cover in every issue (for example, only once in each volume or on inside pages), it is helpful to authors if it is announced on an outside cover where this information can be found. A page reference in the table of contents seems suitable.

(7) American Society for Testing and Materials' CODEN. A Coden is a series of five letters plus a check character, one of which is assigned to every journal. (*CODEN for Periodical Titles* is available from International CODEN Service, c/o Chemical Abstracts Service, Ohio State University, Columbus, Ohio 43210.) The Coden is used by keypunchers (in preference to the full name of the journal) to identify the source of every reference as it is entered into a computerized information-retrieval or data-processing system. Although the keypuncher is, of course, provided with the Coden for each journal, the probability of error on the keypuncher's part is much reduced if each journal bears on its cover the Coden assigned. This is obviously highly desirable for the sake of the future information seeker.

The addition of the Coden to the cover is resisted by editors on two grounds:

(a) It is ugly. Most cover-designers shudder at it, although some of the American Chemical Society journals have succeeded in incorporating it with some semblance of taste. One solution is to provide it on the back cover, the spine, the title page of the issue, or elsewhere easily found.

(b) It may or may not be replaced in the near future by an International Standard Serial Number (ISSN). The Standard for ISSNs has been approved by most members of the Z39 committee of the American National Standards Institute, but the large abstracting services are reluctant to abandon the Coden. (The eight-digit number will almost certainly be even uglier than the Coden.) This should be no real deterrent to editors who decide a Coden should be provided now; if, in a few years, an ISSN is assigned, it may be readily substituted for the Coden.

(8) Announcement of termination, suspension, or change of name. If the journal is to be discontinued, suspended for a known or unknown period, or is to change its name, an announcement on the cover will aid readers, bibliographers, and all secondary services. This information, like all information that has lasting value to users of the journal, should appear in the inside of the journal *as well as* on the cover.

Reference

1. American National Standard for Periodicals: Format and Arrangement (Z39.1-1967), available from American National Standards Institute, 1430 Broadway, New York, New York 10018.

CHAPTER 26

Page identification

In the American National Standard for Periodicals: Format and Arrangement (Z39.1-1967, available from American National Standards Institute, 1430 Broadway, New York, New York 10018), the following recommendation appears:

> 4.2 Each double page spread of a periodical should carry the indications necessary for the rapid identification of the publication. These indications may appear as running heads or footlines, but in principle should always appear in the same place. They should include:
> (1) title of the periodical (abbreviated, if necessary)
> (2) number or the year or date of issue of the volume, preferably both
> (3) page number.

The purpose of the recommendation is to enable those who possess only a fragment of an article—either a single page torn from a reprint or photocopies of two successive pages—to identify the source, and after locating the original material, cite the article correctly. If the running head or foot line employs the standard abbreviation for the title of the periodical (American National Standard Z39.5-1969), users will become familiar with the abbreviation, and again be aided in the direction of correct citation.

Giving the journal's name, the volume, and the year on *every* page would obviously be even more valuable; and it can be done. For years, each page of every journal published by the Elsevier Publishing Company carried both a running head and a running foot line. The running heads contained the authors' names (left-hand page) and short title (right-hand page) besides the page numbers. Each running foot line contained the name of the journal (abbreviated if necessary), volume, year, and inclusive pagination for the article. This last feature requires the only expensive element, since the inclusive page numbers have to be added at a late stage in preparing page proof and checked for accuracy. If the refinement of adding inclusive pagination is omitted, the other information remains constant for every page in the whole volume, and is cheaply provided. An alternative is to print the exact date of publication in the foot line, in which case the foot line remains constant throughout the *issue*. We recommend that one of these alternatives for providing key citation information on every page become standard practice.

Format

CHAPTER 27

Masthead

The masthead of a biomedical journal provides descriptive information, including, if applicable, its name; the owner; publisher, with full address; sponsoring body if other than publisher, with full address; publications committee; editor(s), with full address if other than publisher; editorial board; executive editor or business manager; editorial associates and assistants; place of publication; frequency of issue; subscription price(s); postal notice (within first five pages if second-class rate claimed); purpose and scope of the journal. Most mastheads also contain: addresses to be used by subscribers, authors, advertisers, and notifiers; directions for submission of items for publication (manuscripts, notices); copyright notice; disclaimer for statements appearing in text or advertisements; directions for finding or obtaining Information for Authors; page charges; availability of the journal in microfilm. Some mastheads contain the names of all editorial and executive personnel, date of founding, appropriate colophon, and officers of the controlling professional society.

Mastheads are printed regularly in each issue of the journal and are consistent in position and format. Favored positions are within the first five pages of the issue, next to the table of contents, next to the editorials, or inside the front cover. Whole, one-half, or one-quarter page format may be used.

CHAPTER 28

Placement of features

Placement of the various features in a scientific journal may vary considerably from one editor to another. The format, however, should remain constant over long periods unless there are compelling reasons for change.

For additional discussions regarding placement of certain features, the reader may refer to specific chapters, as follows:

Feature	Chapter
Table of contents	25 and 31
Information for authors	18
Editorials	12
Correspondence	15
Advertisements	11 and 31
Corrections	20

Format

CHAPTER 29

Special features in the final unbound issue of a volume

The final unbound issue of each volume of a journal should include certain information that will be helpful to users. The following items should be considered:

*(1) A title page of the publication should be included, indicating also the volume number, year of issue, place of publication, and, optionally, the editor's name. This page usually becomes the first page in the bound volume (see "Binding Practices").

(2) The reverse of the title page, or a separate page, should carry the masthead, showing the name and address of the editor-in-chief, and the other editors and members of the editorial board when such exist. This principle does not apply to journals that include the masthead within the text of every issue. This page should also include (a) a brief statement of the objectives of the publication, (b) the ownership of the journal and the copyright holder, (c) the number of issues in each volume and frequency of their appearance, (d) address for submission of manuscripts or for information on general editorial matters, (e) subscription price(s), and (f) address for correspondence regarding subscriptions or problems relating to lost or defective copies.

(3) If the journal carries an Information for Authors page in every issue, such a page need not be included as a special feature of the entire volume. If the Information for Authors page is provided only once per volume, its position should be identified in the masthead (see also "Masthead" and "Information for Authors").

*(4) The last numbered page in the issue should be followed by a separate listing of the contents for each issue in the volume, including published

*The pages mentioned in (1) and (4), if intended to be moved to the front of the bound volume, are usually numbered in Roman: i, ii, iii, ...

corrections. These contents should include the exact titles of the articles, authors, and page numbers as they appeared in the individual issues.

(5) An Acknowledgment page is frequently included in the final issue of a volume. Listed are the names of invited special reviewers of manuscripts.

(6) The United States Postal Service now requires each periodical to file and publish a Statement of Ownership, Management, and Circulation. This statement can be carried at the end of a volume.

(7) Indexes for the volume are usually placed last in the issue. Separate author and subject indexes, each arranged alphabetically, are recommended for journals that publish more than 1,000 pages in a volume. In periodicals with fewer pages, the two indexes are often combined. Certain biological journals that publish extensive articles dealing with systematics may include an index to organisms (see "Indexes").

CHAPTER 30

Indexes

Every journal should have one or more indexes to facilitate retrieval of information. An indexer's responsibility is to make the contents of articles in a journal readily accessible under any reasonable heading that may occur to a reader. If the indexer keeps that responsibility uppermost in mind during preparation of the index, he can avoid an arrangement that may facilitate compilation but may complicate or preclude retrieval of published material. Editors should be sure that all index entries are prepared soon after the page proof of each issue is available, so as to keep copy current and avoid a delay when the final issue of a volume goes to the printer.

Many journals publish only a subject index or a combined author-subject index, but for periodicals that cover a wide variety of subjects, contain a large number of papers in one volume, or consist of several sections, it is advisable, and often convenient, to provide more than one index. When special items or sections are a part of a journal, the editor should have the indexer include as much key information about them as is practical in the subject index. Whether separate indexes should be set apart for each major section is a matter to be decided by the editor on the basis of his readers' needs. If separate indexes are provided, informative cross-references should be included in the journal so users will not overlook topics under search. Unfortunately, few, if any, journals that have multiple indexes use full cross-references. Thus, a reader searching for an abstract by Coupling, J. J., will rarely find an entry under Coupling saying "See abstracts." Multiple indexes can, therefore, be a barrier to the retrieval of information rather than an aid. If all categories are intermixed in a single index, it is helpful to have the category (original article, book review, editorial) identified by coding or in some other way. The inclusiveness of the index may depend, at least in part, on whether it is hand-developed or computer-developed.

The most common indexes for a journal are an author index and a subject index. In some cases, however, journals may have additional separate indexes for (1) abstracts of articles that are published in other journals, (2) abstracts of papers that are presented orally at meetings, (3) biologic taxonomy (names of taxa, such as families, genera, species), (4) book reviews, (5) editorials, (6) errata, (7) letters to the editor, (8) obituaries, (9) news columns in which research data are reported, (10) sustaining members or advertisers, and (11) date (day of month) each number is published (see "Publication of Dates for Manuscripts as an Index of Priority"). Indexing of death notices is also useful because this information is otherwise difficult to retrieve. If research data are reported in some detail in a medical news column of a scientific journal, the editor should consider indexing this material as well, to facilitate retrieval by readers. Advertisements are usually not indexed in scientific journals, although they may be in trade publications.

The variety of terms used interchangeably for abbreviated forms of articles creates some confusion—abstract, summary, synopsis, summary-abstract, synopsis-abstract, précis (see "Abstracts"). Indexing of these abbreviated forms, moreover, is not consistent in all journals. In some, abstracts published independently of the corresponding full articles are not indexed at all; such a practice makes it difficult to retrieve information that may not be available elsewhere, as, for example, material presented at, or intended for, oral presentation at a meeting but never published in a form other than the abstract submitted to the program committee. Since such abstracts represent primary publications, they should be indexed. In some journals they are indexed separately, whereas in others they are included in the listing with full-length articles, but are identified by special pagination or in some other way. A separate listing in the index for such abstracts instantly differentiates this type of publication from full-length articles (see "Abstracts"). This type of abstract is to be distinguished from (1) the summary-abstract that accompanies an original article and (2) a section of synopsis-abstracts of material previously published in journals other than the one in which this section appears.

AUTHOR INDEX

An author index usually requires only the transcription of names of all authors of articles and the page references in numerical order. The surname is written first, followed by the given name(s) or initial(s). To avoid confusion for the indexer, librarian, and user, the editor of the journal should try to ensure that the name of an author of multiple articles is always printed in the same way. If John Henry Smith identifies himself in the by-line of one article as John H. Smith, in another as J. H. Smith, and still another as J. Henry Smith, the indexer will probably treat these names as three different authors in the index.

Individual entries for the author index should be typed on cards as soon as page proof for each issue becomes available. The name cards should then be arranged in alphabetical order and stored in a file ready for the printer.

Each time a subsequent issue of a journal becomes available, the new authors' entries need to be inserted alphabetically in the master file.

Questions sometimes arise concerning the arrangement of surnames with prefixes or concerning compound family names. When uncertainties arise about the proper arrangement of names, the system used by libraries or by abstracting, indexing, and citation periodicals should be followed. The author of the article may also be consulted about preferred usage, if this is feasible. In any case, the indexer should try to be consistent in the usage. Diacritical marks and accents on certain letters in names should be retained.

Occasionally, each name in an author index is followed by a subheading consisting of a key subject or phrase to indicate the nature of the topic for each page entry. Such a system may be helpful to users, but limitations of time, space, and costs usually make it impractical. The full citation for a reference is often listed only once, under the first author's name: *Jones, A. B.* Under all co-authors' names, the entry may read: *See Jones, A. B. (first author).* Such entries are not the most convenient for the reader, and the editor may wish to consider repeating the information under each author's name instead of using cross-references.

In indexing book reviews, the editor will need to decide whether to index the author of the book review alone or the author of the book reviewed as well.

SUBJECT INDEX

Ideally, the subject index should be prepared with great care and precision by a person who has knowledge of the subject material in the article. Increasingly, authors are being asked to provide key words or phrases with their manuscripts; these entries are helpful to the person compiling the index, whether the index is prepared entirely by hand or fed into a computer for later print-out as the volume index. Since the selection of key words by authors or by editors who are not skilled in their selection may be so haphazard and inconsistent as to preclude their usefulness in retrieval, authors should be encouraged to use standard terms in their fields, such as those used by Medical Subject Headings (MeSH) or Biological Abstracts Subjects in Context (B.A.S.I.C.) and currently found in biologic and biomedical publications.

The indexer usually places on index cards relevant key words or noun phrases, together with corresponding page numbers in numerical order. The entries may include names of organisms, special chemicals, mathematical expressions, techniques, hypotheses, or even acronyms and symbols. Such words and phrases may be taken from the title, abstract, introduction, main text, or footnotes of each article. The editor can be of great help to the indexer in deciding which key words and standard terms to use, and which sections of the article should be examined in the preparation of the index.

Using words or phrases from the title alone is usually not sufficient for the preparation of a good index. In this connection, the editor need not consider titles submitted by authors to be sacrosanct, but may insert key words in them when necessary to ensure proper indexing. Clever titles, for example, are often provocative, but may obscure the subject of a piece of writing and make proper indexing difficult. Authors should have the opportunity of approving the revised title before the article is set in galley proof.

It is advisable to include cross-references to many entries to accommodate persons with different ideas or reasons for using the index. An editor can help the indexer by making suggestions about cross-references.

VERIFICATION

The editor should insist that index entries be verified against page proof before they are given to the printer. An index is an important and useful part of a journal, and every effort should be made to avoid errors in entries.

CHAPTER 31

Binding practices

Many editors, accustomed to seeing their publications in individual weekly or monthly issues, give little thought to the form in which they are permanently retained in libraries. With few exceptions, the covers and much of the front and back matter of journals owned by libraries are removed before binding. Pages that are not included in the regular numbering sequence are discarded. A few large libraries bind one copy of each journal intact, deliberately preserving the advertisements and other material that may have historical interest at a future date. (Indeed, in a hundred years the advertisements in many journals may be of great interest to social historians, whereas at least some of the text of those journals may be of little interest.)

The general practice being as it is, the editor should be aware that most copies of a journal that are preserved for future reference in libraries will contain none of the material that appears on the covers and none of the advertisements or material interspersed with the advertisements. The tables of contents of individual issues will often thus be discarded from the bound volume; also generally excluded will be the Information for Authors if it appears on a cover, as well as the masthead, which gives the names of the editor and editorial board, the name and address of the publisher, and subscription information.

If advertisements, hotel reservation forms, and other such material appear on the reverse of pages containing scientific information, and if those pages are torn out before binding, this scientific material is obviously lost. For this reason, it is preferable not to print nonscientific material on the reverse of scientific pages and to avoid, whenever possible, interleaving of advertisements with scientific material. Advertisements should be so placed as to allow their easy removal by the binder or others without disturbance of the scientific material or editorial pages that users are likely to want to retain and retrieve.

Some editors believe that some or all of this kind of material should be discarded, since out-of-date information may otherwise be too easily available. An author may, for example, submit his article to the wrong place if he consults the Information for Authors in the bound volume for an earlier year. Other editors believe, however, that some of the nonscientific material should be preserved, as part of the record if for no other reason, and they are unwilling to sacrifice that material merely to protect an author from the naive error of consulting a journal for a previous year when seeking current information.

Each editor should, in any case, be aware of what will and will not be included in the bound volume of his journal and should decide accordingly on the placement of the copy. The format of many journals is intended to encourage the discarding of certain material. The journal is often designed as if it were a book published in parts. Each part (weekly or monthly issue) contains disposable material, such as the covers and front and back matter; when that material is discarded, what remains is a book, paginated continuously from beginning to end and equipped with a table of contents and index for the whole volume and with a new title page designed to be inserted at the beginning. For obvious reasons, the table of contents and index appear in the last issue of the volume; for less obvious reasons, the title page for the volume also often appears in the last issue. All this is designed to permit what many library binders, in fact, do: discard all superfluous matter, move the volume title-page and volume contents-pages to the beginning of the first issue, place the text next, and end with the volume index, in exact imitation of a book. Other library binders discard the superfluous matter, but do not move the volume title-page and volume contents-pages from the last issue, so that the first few pages of the journal (often actually numbered i, ii, iii, . . .) appear at the very end of the bound volume.

The practice of moving the volume title-page and volume contents-pages to the front and of discarding superfluous material is so widespread that the editor who wishes to retain a bound set of his journal in the precise form in which it was published may have great difficulty doing so. The binder, no matter how often he is instructed to bind "as laid," may still move the volume title-page and volume contents-pages to the front, even if he retains the covers and front and back matter. Regardless of the binding practices of libraries, the publisher or editor of a journal would do well to retain at least one complete set in the exact form in which it was published; if that set is bound, it should be stitched and cased, but nothing should be moved from the place in which it originally appeared. The existence of one such set has historical value and could conceivably have value in matters in which legal questions might arise.

APPENDICES

Organizational chart

SCIENTIFIC JOURNAL OFFICE

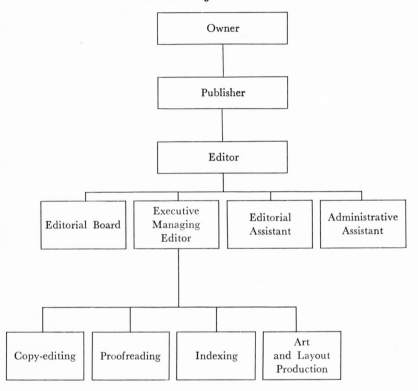

Manuscript flow chart

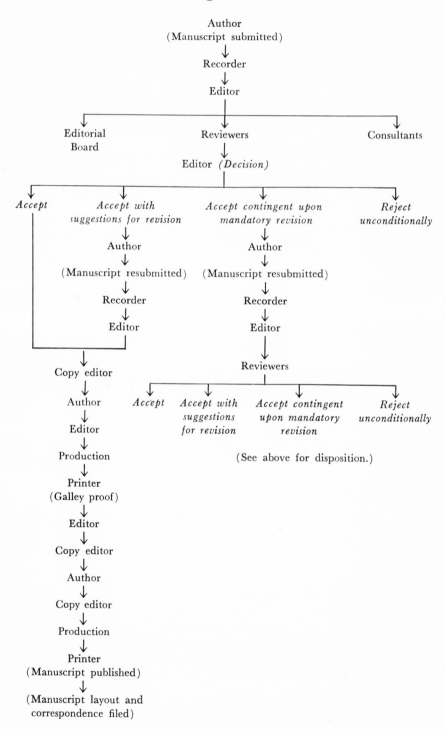

CLINICAL RESEARCH
Abstract Reproduction Form*

TYPE name, address, and telephone number of
author who should receive correspondence in
Box A and complete Box B.

Telephone _____
      (Area code)   office   (Area code)   home

A

Name _____

Address _____

_____

_____

B

Date _____

Payment ($10.00) _____

Check number _____

Purchase order _____

- - - - - - - - - - - - - - TYPE ABSTRACT HERE/BE SURE TO STAY *WITHIN* BORDER - - - - -

CHECK *Preferred Sub-Specialty
      Classification:*

____ Cardiovascular
     Clinical
____   Epidemiology
     Clinical
____   Pharmacology
____ Dermatology
____ Endocrinology*
____ Gastroenterology
____ Genetics
____ Health Care Research
____ Hematology
     Immunology &
____   Conn. Tissue
____ Infectious Disease
____ Metabolism*
____ Oncology
____ Pulmonary
____ Renal & Electrolyte

*Traditionally, *Endocrinology* has
included papers dealing with the
thyroid, adrenal and pituitary
glands, and gonads, while abstracts dealing with the parathy-
roids, calcium and phosphorus
metabolism, bones, thyrocalcito-
nin, diabetes, insulin, glucagon,
and growth hormone have been
considered under *Metabolism*.

PLEASE CHECK ABSTRACT CAREFULLY FOR APPEARANCE BEFORE MAILING

IMPORTANT

The instructions accompanying this form must
be followed COMPLETELY for all abstracts which
are to appear in CLINICAL RESEARCH. Abstracts which do not conform either will be retyped by the publisher at a cost of $15.00 to the
author, or rejected.

Revised June 1975

THIS FORM AS WELL AS THE FORM LETTER
OF TRANSMITTAL MUST BE SIGNED
BY A MEMBER

MEMBER'S SIGNATURE . . . .

*Reproduced by permission of Dr. Robert J. Levine of the American Federation for Clinical
Research.

American Federation for Clinical Research
Rules Applying to Abstracts *

(This abstract form is used by many organizations. Unless explicitly instructed otherwise by officers of these societies, the author should adhere to these rules.)

Please Read This and the Reverse Page Thoroughly Before Preparing Your Abstract

1. Abstracts for CLINICAL RESEARCH must be typed on ABSTRACT REPRODUCTION FORM. Forms will be mailed twice each year to all members, Additional forms may be obtained from: 1. Section Chairmen and Secretaries; 2. C.B. Slack, Inc., 6900 Grove Road, Thorofare, New Jersey 08086.

2. Each abstract must be signed by a member whose name must also be typed on it as either co-author or introducer. One member may sign no more than 2 abstracts for each meeting. One may, however, co-author as many as one chooses.

3. Make TITLE brief, clearly indicating the nature of the investigation. The subject INDEX for CLINICAL RESEARCH is developed only from those words used in the title; choose them carefully. Then state authors' names and institutional affiliations. (If no institution is involved, state city and state.) OMIT degrees, titles, institutional appointments, street address and zip code. Indicate nonmembership by an asterisk, senior membership by two asterisks. Asterisks may not be used as symbols except to connote membership. If no author is a member, type: "(intr. by............)." (See example on next page). Do not mention support of work by a research grant.

4. Organize the body of the abstract as follows:
A statement of the purpose of the study (preferably one sentence).
A statement of the methods used.
A summary of the results presented in sufficient detail to support the conclusions.
A statement of the conclusions reached. (It is not satisfactory to state, "The results will be discussed" or "other data will be presented.")
Do not use subtitles, e.g., methods, results.

5. Simple tables or graphs—neat and in black ink—may be included if they fit within form rectangle.

6. Use of standard abbreviations is required. Abbreviations accepted by this journal are listed in two places: 1) Instructions to Authors, J. Biol. Chem. 246:1-8, 1971, and 2) *Stylebook:* Editorial Manual, Fifth Edition, American Medical Association, Chicago, 1971. Acceptable abbreviations may be used without definition if the author chooses. Abbreviations not defined in the two cited references must be defined by placing them in parentheses after the full word the first time they appear. Some listed abbreviations may be unfamiliar to large numbers of readers; at the author's discretion these also should be defined similarly. ABBREVIATIONS SHOULD BE AVOIDED IN THE TITLES OF ABSTRACTS.
Use numerals to indicate numbers except to begin sentences.

7. *Nonproprietary* (generic) names are required the first time a drug is mentioned, written in small letters. *Proprietary* names are always *capitalized*, e.g., acetazolamide (Diamox).

8. Use type no larger than in examples. Do not use type that simulates script. Use carbon ribbon or slightly used black silk ribbon. (Brand-new ribbons smudge, old ones

print too faintly.) PRACTICE typing the abstract in a rectangle 6 x 4¾" before using this form.

9. DO NOT ERASE. Abstract will appear in CLINICAL RESEARCH exactly as submitted. Abstracts with smudges, errors, misspellings, poor hyphenations, skipped lines, typed-in margins, incorrect abbreviations, too faint typing, etc. (or not conforming to prescribed rules) may require retyping by the publisher at the author's expense.

10. CAPITALIZE entire TITLE. Underline authors' names (Underlining or capitalization for emphasis in text is unacceptable.) Single-space all typing (no space between title and body or between paragraphs). Indent each paragraph 3 spaces. Do not indent title. Draw special symbols in black ink.

11. A) Provide original plus 12 copies of the completed abstract reproduction form (Xerox or equivalent).
B) The enclosed form letter of transmittal must be completed and signed by a member. Provide four copies of this form letter. If more than one abstract is submitted, a separate letter of transmittal must accompany each abstract.

12. A) No paper may be presented at the National Meeting if it has been either published or presented at any other national meeting prior to the date of presentation.
B) AFCR papers may be presented only by authors who have not passed their forty-first birthdays; these may be active members or introduced nonmembers.

13. SUBMIT abstract material for:
a. NATIONAL MEETING to: Publisher of
CLINICAL RESEARCH
c/o Charles B. Slack, Inc.
6900 Grove Rd.
Thorofare, N.J. 08086
b. REGIONAL MEETING to: *Section Chairman* (Eastern, Western, Southern Mid-Western)
Adhere to published mailing deadline; ABSTRACTS MAY NOT BE WITHDRAWN later than one week after mailing deadline. DO NOT FOLD this sheet. MAIL abstract (12 copies, Xerox or equivalent), form letter of transmittal (4 copies), payment (see rule 14), and (if receipt is required) a self-addressed postal card. Use cardboard backing. Any material (supporting data or illustrations) enclosed other than that specified will be discarded. Use FIRST CLASS; if distance is over 400 miles, use AIR MAIL. Abstracts are not acceptable if postmarked after the mailing date. *It will be impossible to consider abstracts received more than five days after the deadline.*

14. Payment of $10.00 (check made out to AFCR) must accompany all abstracts submitted for consideration for all Sectional and National meetings of the AFCR, except:
a. When institutional requirements dictate that a purchase order must be used; if so type the number and date of purchase order on the form where indicated.
b. When form is used to transmit abstract to an organization other than the AFCR which publishes in CLINICAL RESEARCH.

*Reproduced by permission of Dr. Robert J. Levine of the American Federation for Clinical Research.

CLINICAL RESEARCH
SECRETARY CHECKLIST *

Before mailing, please check your abstracts for the following common errors:

_____ Number of enclosed check and/or purchase order must be stated in Box B.

_____ Be certain TITLE IS COMPLETELY CAPITALIZED

_____ Author's and introducer's names must be continuously underlined

_____ Initials or first names must precede last name

_____ Institutional affiliation and city are required

_____ Be sure that street address, zip code, degrees and grant support are NOT listed in abstracts

_____ Do not begin author's name or address on a new line unless necessary (see example)

_____ Do not indent title

_____ Begin body of abstract on a new line which is indented three spaces

_____ Abstract must stay within rectangular borders

_____ Do NOT blacken borders of abstract rectangle

_____ Left hand border must be perfectly straight

_____ Do NOT squeeze letters or lines

_____ Smudges or faint typing may require retyping of the abstract

_____ There must be no unacceptable abbreviations (Rule 6)

_____ Do NOT use CAPITALS or UNDERLINE for emphasis

_____ Check accuracy of SPELLING and HYPHENATION

_____ There should be no space between lines or paragraphs

_____ A conclusion must be stated. A hopeful promise of additional data or discussion is not acceptable

_____ The form must be signed by a member

_____ Be sure asterisks are used to designate nonmembers* and senior members**

_____ Asterisks are NOT to be used as reference symbols except to connote membership status

_____ Be certain to designate subspecialty classification

ENCLOSURES

_____ $10.00 abstract payment for National or Regional AFCR Meeting (See Rule 14).

_____ Original plus 12 copies (Xerox or equivalent) of the abstract form.

_____ Form letter of transmittal (4 copies, Rule 11B)

_____ Self-addressed postal card (optional, if acknowledgment of receipt is required).

(Other enclosures will be discarded—Rule 13)

EXAMPLES

```
ARRHYTHMIAS OF THE HEART: MECHANISM(S). A. C. Brown,* J. B. Green,*
and R. T. White* (intr. by W. W. Sloane). Department of Medicine,
Yale University, New Haven, Connecticut.
    Digitalis (D), potassium (K+), and nicotine (N) induce automaticity
and propagation block. The initial event is enhanced conduction (C).
```

```
EFFECTS OF VOLUME LOADING ON RIGHT VENTRICULAR VOLUME AND TRICUSPID
REGURGITATION IN ACUTE PULMONARY HYPERTENSION. A.B. Wright,* D.F.
Brown, and J. W. Thomas,** Department of Medicine, University of West
Virginia, Morgantown, West Virginia.
    Right ventricular volume (RVV), function, and the development of
tricuspid regurgitation (TR) determine cardiac output (CO) in acute
```

*Reproduced by permission of Dr. Robert J. Levine of the American Federation for Clinical Research.

Index

Regional editors, 6
Register, Federal, 29, 81
Register of Copyrights Report, 80
Rejection of
  article, reasons for, 16
  paper, responsibility for, 19
  solicited item, 58
  symposium, 61
Rejection, responsibility for, 11, 19
Remuneration of reviewers, 5, 14, 64, 66
Reprints, 73, 74
Reproduction of
  abstract, sample form, 113
  journal contents, 73, 75, 79, 80
Republication, abuse, 37
Reputation of
  author, 33
  editor, 4
Research
  article, solicited, 58
  ethical assessment of, 32
  journal
    functions, 3
    purpose, 1
Responsibility (*see also* specific roles and
    functions)
  acceptance-rejection decisions, 11, 19, 31
  administrative, for journal, 4
  advertisements, 50
  author's, for references cited, 89-90
  by-lines, 40-41
  decision to publish, of editors and re-
    viewers, 1
  delegation by editor, 6
  editor's, 4, 11, 19, 31, 108
  indexer's, 105
  proofreading, 27, 28, 92
  science writer's, 39
Retrieval of information, 103
Review(s) (*see also* Reviewers; Review-
    ing)
  books, 58, 64, 104, 105
  delayed, 42
  journal, functions, 3
  manuscript
    criteria for, 15, 20
    technical and administrative aspects,
      24-28
  negative, 22, 23
  peer, advantages, 38
  prejudiced, editorial handling of, 12,
    13, 22, 66
  process for, 11-17, 112
  second negative evaluation, 23
  signed, 12, 13, 64
  solicited articles, 58
  third, need for, 26

Review article, 58
Review boards, institutional, 29
Review journal, 3
Reviewers (*see also* Book reviewers; Re-
    views; Reviewing)
  anonymity, 12, 13, 17
  attitude toward manuscript under re-
    view, 18
  biased, editorial handling, 12, 13, 22,
    66
  caveats, 60
  comments
    aid to author, 15
    grammar and, 21
    publication of, 60
    role in editorial decisions, 21
    style and, 21
    transmission to author, 16, 20
  compensation, 5, 14
  confidentiality, 15
    letter to editor, 19
  contradictory, 12
  criticisms by, usefulness to editor, 20
  definition, 11
  editorial board and, 6, 112
  editorial communication with, 17, 26
  fallibility of, 26
  feedback from editor regarding ultimate
    decision, 17
  file, 25
  function, 91
  grading of manuscripts by, 14
  guidance by editor, 14
  guidelines for, 14, 18-21, 29-35, 65-66
  identification, 12
  inexperienced, 14, 19
  information about disposition of paper,
    17
  letter to editor, 60
  listing on acknowledgment page, 102
  negative evaluation by, 22, 23
  optimum number, 11, 12
  performance, editorial evaluation, 25
  prejudiced, 12, 13, 22, 66
  protection of unpublished manuscripts,
    18
  publication of names, 25
  recommended by editorial board, 5
  remuneration, 5, 14
  responsibility for disposition of paper,
    1, 19
  selection of, 5
  suggestions for revision, 20
  third, 26
Reviewing, 11-35 (*see also* Reviews; Re-
    viewers)
  acceptance-rejection decisions, 11